T0302275

"*Heroes of Environmental Diplomacy* tells an important story. The individuals chronicled here took on daunting challenges in negotiating global agreements on some of the biggest issues of our times. Through courage and determination, they managed to forge positive action. These stories and their collective message reaffirm one of my core beliefs – multilateral diplomacy, when carried out with vision and persistence, actually works. The efforts and achievements highlighted in these chapters are a clear illustration of three crucial action points for achieving the 2030 Agenda for Sustainable Development. First, every human being, especially our leaders, must be unrelenting in their efforts to do the needful in their sphere of influence. Second, we must never lose hope in the power of the individual to effect transformational change. Third, we can and must find ways to work together, even when it seems impossible to find common ground, to solve these extremely complex sustainability challenges. Let us heed these lessons that have been gifted to us by these environmental heroes."

Amina J. Mohammed, *United Nations Deputy Secretary-General*

"*Heroes of Environmental Diplomacy* is a fascinating read. While many books focus just on the issues, this one provides valuable insights into the people who made things happen. As someone who has been involved in the UN and intergovernmental negotiations for many years, I can attest to the importance of unique individuals in helping secure success. As we enter a critical decade for humanity in our struggle with both climate change and biodiversity, this book shows we can succeed in crafting a path to a more sustainable future and illustrates the role individual heroes can play."

Elizabeth Maruma Mrema, *Executive Secretary of the UN Environment Programme's Convention on Biological Diversity*

"Tina Turner couldn't have been more wrong when she sang 'We don't need another hero'. As this book shows, humanity has always needed heroes to prevent short-termist, populist and destructive forces holding sway and leading the charge towards environmental ruin. Who will step forward and be the heroes of this decade? Heaven knows they're sorely needed – and this book shows with both depth and colour what heroism entails and what it brings."

Richard Black, *Honorary Research Fellow, Grantham Institute, Imperial College, former BBC Environment Correspondent*

"*Heroes of Environmental Diplomacy* opens with this simple question 'What makes a hero?' The answers are found woven in the inspiring stories of the courageous, ordinary people profiled in this treasure of a storybook. Readers will enjoy an intimate look at the struggles, perseverance, hope, and stamina that inspired the people highlighted in the book to work tirelessly to turn aspirations into realities. This treasure of a book will greatly expand the knowledge base of the foreign policy community, but equally important it will inspire a whole new generation of justice-seekers to keep on talking, writing, debating, compromising – and ultimately creating the future agreements needed to secure the brighter future these ordinary heroes have gifted us the opportunity to pursue."

Mark Ritchie, *Global Minnesota*

"*Heroes of Environmental Diplomacy* makes a profound contribution to the literature on global governance innovation by detailing the practical, creative, and often-times courageous ways diplomats have navigated complex and politically fraught multilateral environmental negotiations. With original insights on circumventing spoilers and mobilizing multi-actor coalitions for progressive environmental change, the book will serve as an important playbook for policy researchers, practitioners, and advocates concerned with updating global institutions, legal instruments, and other arrangements to address current and emerging global challenges."

Dr. Richard Ponzio, *Senior Fellow and Director, Global Governance, Justice & Security Program, Stimson Center*

"In Greek mythology, the hero was half a God. The etymology remains but the meaning has evolved and become more varied. A hero has become a sort of superwoman/superman and protector, a brave and courageous person, in brief, a winner. In French a héros can be the main protagonist of a narrative. Their trait of character is to be embodied in an ideal system of values, to show self-sacrifice in front of critics and perseverance in a world that is often disheartening. Their feats and achievements must be reported to be models of excellence. This is the merit of the present publication.

During my career, I had the privilege to meet several of the heroes described within, who deserve to be mentioned. The French author Malraux wrote: 'There is no hero, without an audience'. I would say that they have mostly been convincing leaders. Indeed, they were seasoned negotiators and brilliant spokespersons, supported, however, by dedicated assistants conducting specific discussions. Corridor diplomacy is essential in the multilateral arena. Therefore, the whole cast are heroes!"

Marcel A. Boisard, PhD, Former *UN Assistant Secretary General and Executive Director of UNITAR*

"History is not the result of anonymous or mechanic forces, but of human action. And human action, while shaped by environmental and social factors, essentially means that human leadership is important to shape history's course. This has sometimes been overlooked, and that is especially true for environmental history, since here the natural focus is exactly on environmental factors and social activities as aggregates more than individual action. But also environmental protection has a deeply human side and it is good that the new book edited by Felix Dodds and Chris Spence, *Heroes of Environmental Diplomacy: Profiles in Courage*, brings to the forefront those who have fought on the international stage for rules making the world more a fair, equitable and sustainable for all of us to live in together. It is interesting to see a common thread for the chapters, namely that not only professionalism and success, but also the human side and compassion of the actors is important for their career. Regardless of their focus, more technical ones like the 'Basel Convention on the Control of Transboundary Movements of Hazardous Wastes and their Disposal' or more conservation-driven once, environmental conventions, treaties and events like the Ramsar Convention on Wetlands, the International Whaling Commission or the Rio Earth summit, show that environmental heroes need professionalism and compassion. This book is a welcome addition to the existing, more technical literature on environmental history well suited for students, academics and people working in the field of international environmental cooperation."

Dr. Bernhard Seliger, *Resident Representative Hanns Seidel Foundation – Korea Office*

"I have read this book with pleasure. The book concept is refreshing and gives a fascinating personalized insight into how important steps on global environmental progress have been achieved. The authors manage to write the various stories in a way which is highly readable and give great background about serious issues and at times amusing characterizations of people who have been global environmental heroes. Definitely recommended reading for those interested in the environment and the way international negotiations proceed."

Rob Visser, *former Acting Director of the Environment Directorate, OECD*

"*Heroes of Environmental Diplomacy* vividly illustrates how leadership arises from necessity and how individuals with vision and stubborn optimism make diplomacy and multilateralism work and bring lasting results for human development. This is a very accessible book for anyone interested to learn about how some of the most important struggles in multilateral environmental history have been won, and the inspiring individuals at the center of those achievements."

Tom Rivett-Carnac, *Founder, Global Optimism*

HEROES OF ENVIRONMENTAL DIPLOMACY

Today more than ever, when the world is beset by environmental, social, health-care and economic challenges, we need courage in our politics, both nation-ally and globally. This book tells the stories, some for the first time, of twelve individuals who made heroic contributions to protecting our planet through ground-breaking international treaties.

Can individuals change the world? Today, when impersonal forces and new technologies seem to be directing our lives and even our entire planet in ways we cannot control, this question feels more relevant than ever before. This book argues that we can all make a difference. It tells inspiring stories of individuals who have had a global impact that is beyond dispute, as well as others who have brought about change that is understated or hard to measure, where the scale of the impact will only become clear in years to come. While some are scientists, others are politicians, diplomats, activists, and even businesspeople. However, they all share the qualities of perseverance, patience, a willingness to innovate or try new approaches, and the endurance to continue over years, even decades, to pursue their goal. Drawing on interviews and the inside stories of those involved, each chapter follows one or more of these heroic individuals, a list which includes Luc Hoffmann, Mostafa Tolba, Maria Luiza Ribeiro Viotti, Raul Oyuela Es-trada, Barack Obama and Paula Caballero.

Presenting an uplifting and gripping narrative, this book is an invaluable re-source for students, scholars, activists and professionals who are seeking to un-derstand how consensus is reached in these global meetings and how individuals can have a genuine impact on preserving our planet and reinforcing the positive message that global cooperation can actually work.

Felix Dodds is an Adjunct Professor at the University of North Carolina (UNC) and an Associate Fellow at the Tellus Institute. He was the co-director of the 2014 and 2018 Nexus Conference on Water, Food, Energy and Climate. In 2019, he was a candidate for the Executive Director of the United Nations Environment Programme (UNEP). He is the UNC lead PI on the Belmont Forum grant on Disaster Risk Reduction and Resilience. He has written or edited over 21 books, and his last book was *Tomorrow's People and New Technologies: Changing How We Live Our Lives*. His other books have included the *Vienna Café Trilogy* which chronicles sustainable development at the international level. He co-wrote the first *Only One Earth* with the father of Sustainable Development Maurice Strong and Michael Strauss, the second *From Rio+20 to the New Development Agenda* with Jorge Laguna Celis and Ambassador Liz Thompson and the last one *Negotiating the Sustainable Development Goals* with Ambassador David Donoghue and Jimena Leiva Roesch. Felix was the Executive Director of Stakeholder Forum for a Sustainable Future from 1992 to 2012. He played a significant role in promoting multi-stakeholder dialogues at the United Nations and proposed to the UN General Assembly the introduction of stakeholder dialogue sessions at the United Nations Commission on Sustainable Development. In 2011, Felix was listed as one of 25 environmentalists ahead of his time. Also, in 2011 he chaired the United Nations DPI 64th NGO conference – 'Sustainable Societies Responsive Citizens' which put forward the first set of indicative Sustainable Development Goals. From 1997 to 2001, he co-chaired the UN Commission on Sustainable Development NGO Steering Committee.

Chris Spence is a writer and environmentalist. He has worked internationally on sustainable development, conservation, climate change and health policy, and he held leadership positions at non-profit organizations in New York, New Zealand and California. He also consulted widely for the United Nations, IUCN-World Conservation Union and the International Institute for Sustainable Development (IISD), and he has undertaken assignments in more than 40 countries on five continents, focused, in particular, on climate change and sustainable development policy and practice, as well as international law. He has also served as a political advisor and journalist and been on the boards of several environmental organizations. An award-winning writer, Chris is the author or co-author of several books, including *Global Warming: Personal Solutions for a Healthy Planet* (Palgrave Macmillan, 2005) and *Rock Happy* (2021). He holds MA (Hons) and BA degrees in Political Science and History from Victoria University. Originally from the United Kingdom, Chris lived in New Zealand, New York, and the San Francisco Bay Area before moving to Ireland in 2020, where he currently lives with his family.

HEROES OF ENVIRONMENTAL DIPLOMACY

Profiles in Courage

Edited by
Felix Dodds and Chris Spence

Routledge
Taylor & Francis Group
LONDON AND NEW YORK

earthscan
from Routledge

Cover image: Getty Images

First published 2022
by Routledge
4 Park Square, Milton Park, Abingdon, Oxon OX14 4RN

and by Routledge
605 Third Avenue, New York, NY 10158

Routledge is an imprint of the Taylor & Francis Group, an informa business

British Library Cataloguing-in-Publication Data
A catalogue record for this book is available from the British Library

Library of Congress Cataloging-in-Publication Data
A catalog record has been requested for this book

ISBN: 978-1-032-06547-2 (hbk)
ISBN: 978-1-032-06544-1 (pbk)
ISBN: 978-1-003-20274-5 (ebk)

DOI: 10.4324/9781003202745

Typeset in Bembo
by codeMantra

This book is dedicated to all those who have taken part in intergovernmental negotiations who have been trying to secure a fair, equitable and sustainable world for all of us to live in together.

CONTENTS

FIGURES

CONTRIBUTORS

Felix Dodds is an Adjunct Professor at the University of North Carolina (UNC) and an Associate Fellow at the Tellus Institute. He was the co-director of the 2014 and 2018 Nexus Conference on Water, Food, Energy and Climate. In 2019, he was a candidate for the Executive Director of the United Nations Environment Programme (UNEP). He is the UNC lead PI on the Belmont Forum grant on Disaster Risk Reduction and Resilience. He has written or edited over 21 books, and his last book was *Tomorrow's People and New Technologies: Changing How We Live Our Lives*. His other books have included the *Vienna Café Trilogy* which chronicles sustainable development at the international level. He co-wrote the first *Only One Earth* with the father of Sustainable Development Maurice Strong and Michael Strauss, the second *From Rio+20 to the New Development Agenda* with Jorge Laguna Celis and Ambassador Liz Thompson and the last one *Negotiating the Sustainable Development Goals* with Ambassador David Donoghue and Jimena Leiva Roesch. Felix was the Executive Director of Stakeholder Forum for a Sustainable Future from 1992 to 2012. He played a significant role in promoting multi-stakeholder dialogues at the United Nations and proposed to the UN General Assembly the introduction of stakeholder dialogue sessions at the United Nations Commission on Sustainable Development. In 2011, Felix was listed as one of 25 environmentalists ahead of his time. Also, in 2011 he chaired the United Nations DPI 64th NGO conference – 'Sustainable Societies Responsive Citizens' which put forward the first set of indicative Sustainable Development Goals. From 1997 to 2001, he co-chaired the UN Commission on Sustainable Development NGO Steering Committee.

Chris Spence is a writer and environmentalist. He has worked internationally on sustainable development, conservation, climate change and health policy, and held leadership positions at non-profit organizations in New York, New Zealand and California. He also consulted widely for the United Nations, IUCN-World Conservation Union and the International Institute for Sustainable Development (IISD), and has undertaken assignments in more than 40 countries on five continents, focused, in particular, on climate change and sustainable development policy and practice, as well as international law. He has also served as a political advisor and journalist and been on the boards of several environmental organizations. An award-winning writer, Chris is the author or co-author of several books, including *Global Warming: Personal Solutions for a Healthy Planet* (Palgrave Macmillan, 2005) and *Rock Happy* (2021). He holds MA (Hons) and BA degrees in Political Science and History from Victoria University. Originally from the United Kingdom, Chris lived in New Zealand, New York, and the San Francisco Bay Area before moving to Ireland in 2020, where he currently lives with his family.

Craig Boljkovac is an expert in the field of international chemicals and waste management, based in Geneva, Switzerland. He works with the School of Environment, Tsinghua University; and Regional Centres (Asia-Pacific) for the Basel and Stockholm Conventions. He also acts as a Special Advisor for Environmental Affairs for GEM (Green Eco-Manufacture), a world-leading urban mining and electric vehicle battery materials company. He also served as a Senior Consultant to UNEP, UNIDO, the Government of Switzerland, and the European Union's MUTRAP office (European Trade Policy and Investment Support Project) in Hanoi, Vietnam. Craig has advised the Canadian and Swiss Governments, and the European Commission on international environmental issues.

He previously worked with the United Nations Institute for Training and Research (UNITAR), including as Manager of its Chemicals and Waste Management Programme (CWM). Upon his departure, CWM managed projects in over 100 countries. Under his leadership, UNITAR initiated (with OECD) a new, innovative global work area on Nanotechnology – the first such programme in the UN system.

Craig was Chair of the IOMC (Inter-Organization Programme for the Sound Management of Chemicals), which is the pre-eminent international coordinating mechanism in the field of chemicals and waste management, comprising the programme heads of nine international organizations.

Before joining the United Nations, Craig spent the balance of his career working in organizations in Canada at the national and international levels.

Joanna Depledge is Fellow at the Cambridge Centre for Environment, Energy and Natural Resource Governance (CEENRG). She has been following the climate change negotiations and global environmental politics more broadly, for more than 25 years, as a staff member/consultant at the UN Climate Change Secretariat (1996–2002), a reporter for the *Earth Negotiations Bulletin* (1999–2003) and an academic researcher affiliated with Cambridge University (2003 to date). Up to the end of 2020, Joanna was also Editor of the international journal *Climate Policy* and is now a member of its editorial board. Joanna is a founding member of Cambridge Zero, a member of the research network Climate Strategies, and sits on the steering committee of the UNEP Production Gap Report. In 2017, she curated 'The UNFCCC Story' a unique exhibition at the UN Climate Change Secretariat's headquarters documenting the history of the climate change negotiations. Joanna has published widely on global environmental negotiations and politics, with a focus on climate change. She holds a PhD from University College London.

Luiz de Andrade Filho is a career diplomat since 2012. In the Ministry of Foreign Affairs of Brazil, he has held positions in the Investment Division and the Climate Change Division. He was part of the climate change negotiations team in the Paris, Marrakesh, Bonn-Fiji, Katowice, and Madrid-Chile Conferences of the Parties to the UN Framework Convention on Climate Change (2015–2019). Before joining Brazil's foreign service, he had worked on sustainability issues in Brasília-based organizations for eight years. He holds a BSc in International Relations from the University of Brasilia and a MPhil in Development Studies from the University of Cambridge.

Andrew Higham is a strategist who has worked across diverse sectors, policy domains and all levels of government for two decades, most recently in shaping international climate change policy. Within the UNFCCC from 2011 to 2016, Andrew drove the strategy for reaching a universal, legally binding agreement. He provided the strategic and negotiating support to South Africa in 2011 and drafted the Durban Mandate at COP 17, was the Secretary to the Ad Hoc Working Group on the Durban Platform during its first two years, and then managed the large drafting and advisory team that developed and delivered the Paris Agreement. In 2016, he was invited by the UNFCCC Executive Secretary to take on the role of Chief Executive of Mission 2020 and focused on driving down global emissions at the pace and scale required by science for full decarbonisation by 2050.

Through his leadership, the Mission 2020 campaign has been credited with many achievements across all areas of global climate action. In 2008–2010, he designed and delivered by COP 16 the Technology Mechanism under the Convention and undertook research designing financial instruments to support international climate action, including precursors for the Green Climate Fund. He is presently Special Advisor to the UNFCCC High-Level Climate Champions, Director of the Future of Climate Cooperation initiative, and Visiting Fellow at Oxford University, a partner of Global Optimism, Strategic Advisor to the Earth4All initiative, and Managing Director of Plexus Strategy. He is also a member of the Urban Leadership Council.

André Aranha Corrêa do Lago is a career diplomat since 1983. He is currently the Ambassador of Brazil to India and Bhutan (since 2018) and was previously the Ambassador to Japan (2013–2018). He has held positions in Brasília in areas such as Trade Promotion, International Organizations, Protocol, Sustainable Development, Energy and Climate Change. He was the Director-General of the Department of Energy of the Ministry of Foreign Affairs, between 2008 and 2011, and Director-General of the Department of the Environment, between 2011 and 2013. He was also Brazil's chief negotiator for climate change (2011–2013) and for Rio+20 (2012) and sherpa of the High-Level Panel on Global Sustainability. He has served at the Brazilian Embassies in Madrid (1986–1988), Prague (1988–1991), Washington DC (1996–1999) and Buenos Aires (1999–2001), and at the Mission to the European Union in Brussels (2005–2008). He holds a BSc in Economics from the Federal University of Rio de Janeiro.

Patrick Ramage learned from and collaborated with Sidney Holt for more than a quarter century and says it was one of the greatest privileges of his life. Patrick began his career in environmental advocacy in 1992 as director of GLOBE USA, the U.S. branch of the Global Legislators Organization for a Balanced Environment (GLOBE). Patrick worked closely with then-U.S. Senator John Kerry and a bipartisan group of legislators from the United States House and Senate committed to international action on pressing environmental challenges. In his subsequent work with the International Fund for Animal Welfare (IFAW), Patrick has been at the forefront of worldwide efforts to reduce ocean noise, mitigate disruptive shipping routes, and end commercial whaling by Japan, Iceland, and Norway. By maintaining strategic partnerships and open dialogues with leading scientists, government officials, industry executives, media representatives, mariners, fishermen, and shipping organizations, Patrick is advancing practical solutions to the most urgent threats to whales and the global ocean.

Patrick has led more than a dozen IFAW delegations to the International Whaling Commission, served as NGO representative to the Arctic Council, the World Trade Organization, and the UN Conference on Environment and Development (UNCED), and the UN Conference on Population and Development. Early in his career, Patrick served as a German and Russian linguist with U.S. Army Military Intelligence and completed Survival, Evasion, Resistance to Interrogation and Escape (SERE) training with the British Special Air Service (SAS).

Izabella Teixeira is Co-Chair of the UN Environment Programme's International Resource Panel. She is an expert in environmental management, impact assessment and licensing. She is the former Environment Minister of Brazil (2010–2015) and was nominated a Champion of the Earth in 2013. From 2008 to 2010, she was the Deputy Minister of the Environment of Brazil. A career public servant, Ms Teixeira held positions at the Brazilian Institute for the Environment and Natural Resources and the State Government of Rio de Janeiro. She has also been the head of the Brazilian Delegation in the Cancun, Durban, Doha, Warsaw, and Paris Conferences of the Parties to the UN Framework Convention on Climate Change (2010–2015). At the invitation of UN's Secretary General, she was a member of the High-Level Panel on Global Sustainability. She was also a key leader of the 2012 UN's Rio+20 Conference on Sustainable Development. After Rio+20, she was again appointed by the UN's Secretary-General as a member of the High-Level Panel on the Post-2015 Development Agenda. She holds a BSc in Biological Sciences from the University of Brasília, and an MSc and a PhD in Energy Planning from the Federal University of Rio de Janeiro.

Irena Zubcevic is Director at the Policy and Review Branch at the United Nations Department of Economic and Social Affairs. She has over 20 years of experience in international development, international relations, diplomacy, policy, advocacy, and project management. As a delegate in the Permanent Mission of Croatia to the United Nations in New York, she was engaged with many sustainable development processes, including supporting Croatia's Presidency of the UN Economic and Social Council, being a Vice-Chair of the Commission on Sustainable Development (CSD), Vice-Chair of the Economic and Financing Committee of the General Assembly, and a Vice-Chair of the Bureau for the Small Island Developing States' International Meeting in Mauritius. She was a negotiator for her country at the 2005 World Summit and member of the first generation of delegates in the Peacebuilding Commission. Working at the United Nations Secretariat since 2008, she was supporting CSD, Rio+20 Conference and post-2015 development agenda negotiations. Currently, she is supporting the work of the high-level political forum on sustainable development (HLPF) and, in particular, Voluntary National Reviews (VNRs) that countries present at the HLPF every year to show their national implementation of the 2030 agenda for sustainable development and sustainable development goals. In this role, she has been working with high officials in governments and outside governments, civil society, academia, scientific community as well as civil society organizations. She is an organizational innovator and agile leader with an entrepreneurial mindset and is thriving in high-paced diverse environments. She gives talks and lectures on sustainable development to various international forums and meetings.

FOREWORD

In 1972 at the Stockholm Conference on the Human Environment, Maurice Strong, Secretary-General of the Conference who was later to become the first Executive Director of the United Nations Environment Programme or UNEP, noted that "A point has been reached in history when we must shape our actions throughout the world with a more prudent care for their environmental consequences." The year 1972 was to become the anchor of environmental governance both within and amongst countries.

Within a decade of the Stockholm Conference, we saw the beginnings of environmental ministries being set up across the world. We saw the perspective of developing countries find a strong voice as we embarked on a fundamental shift in global environmentalism. We saw the emergence of irrefutable science that has consistently sounded the alarm on the state of the planet, and therefore people's health. We saw greater environmental awareness. We saw countries come together, underlining the interdependence of challenges, to craft a slew of multilateral environmental agreements that together weave a tapestry of global commitment to the one earth we call home.

As this book outlines, in every decade of environmental milestones, we have seen environmental leadership for the planet and for people. The profiles in this book represent vision, courage and determination. As the environmental crises before us multiply, these leaders serve as a powerful source of inspiration, that it is indeed possible for us to overcome the many obstacles that stand in the way of living in harmony with nature.

And live in harmony with nature we must, because a healthy environment will make the world a better place – for everyone, everywhere. Today, we live in a time of unprecedented crisis. The triple planetary crisis of climate change, of biodiversity loss and pollution and waste threaten to undermine decades of progress in poverty reduction and our efforts to achieve the sustainable development goals.

The challenges we face are so acute, no one country, or group of people can go it alone. We need everyone on board to make the kinds of transitions that are critical to making peace with nature. The good news is that we are seeing this happen. Environmental leaders over the next 50 years are today's youth reminding us that it is our children and grandchildren who stand to inherit a destroyed earth. Environmental leaders are businesses that move beyond paying lip service to sustainability and investors that see through greenwashing.

Environmental leaders are political leaders that ensure a just transition away from fossil fuels. Environmental leaders are indigenous leaders and local communities who are the keepers of our natural assets. Environmental leaders are economists who redefine economic growth to factor in nature. Environmental leaders are judges and jurists, who today with foresight, enshrine and uphold the human right to a healthy environment. Environmental leaders are the multilateral systems that are inclusive and networked, committed to leaving no one behind. Environmental leadership for the planet will succeed because we stand on the shoulders of the remarkable giants outlined in this book.

Former Indian Prime Minister Indira Gandhi, at the Stockholm conference on the human environment noted, "We are gathered here under the aegis of the United Nations. We are supposed to belong to the same family sharing common traits and impelled by the same basic desires, yet we inhabit a divided world." Environmental diplomacy is our chance to demonstrate the solidarity and vision that a divided world so desperately needs.

Inger Andersen
UN Under-Secretary-General
and Executive Director of the United Nations Environment Programme

ABBREVIATIONS

AGBM	Ad Hoc Group on the Berlin Mandate
ALBA	Alianza Bolivariana para los Pueblos de Nuestra América
AMR	Annual Ministerial Review
AOSIS	Alliance of Small Island States
ASEAN	Association of Southeast Asian Nations
BAN	Basel Action Network
BASD	Business Action for Sustainable Development
BASIC	Brazil, South Africa, India and China
BPOA	Barbados Plan of Action
BRICS	Brazil, Russia, India, China, South Africa
CBD	Convention on Biological Diversity
CDM	Clean Development Mechanism
CEB	Chief Executive Board
CEPCIL	Committee of Experts on the Progressive Codification of International Law
CFC	chlorofluorocarbon
CHS	*see* UNCHS
CLI	Country-Led Initiative
CMS	Convention on Migratory Species
Cof3	Committee of Three
COP	Conference of Parties (of the UNFCCC)
CSD	*see* UNCSD
CSR	Corporate Social Responsibility
DCF	Development Cooperation Forum
DDA	Doha Development Agenda
DESA	*see* UNDESA

DPCSD	Department for Policy Coordination and Sustainable Development (United Nations)
DSD	Division on Sustainable Development (United Nations)
EAC	East Africa Community
ECOSOC	Economic and Social Council (United Nations)
EEZ	Exclusive Economic Zone
EMG	Environment Management Group
ENB	Earth Negotiations Bulletin
ESM	environmentally sound management
FAO	Food and Agriculture Organization (United Nations)
FCCC	*see* UNFCCC
FDI	foreign direct investment
G7	Group of Seven industrialized countries (Canada, France, Germany, Italy, Japan, UK, USA)
G20	Argentina, Australia, Brazil, Canada, China, France, Germany, India, Indonesia, Italy, Japan, the Republic of Korea, Mexico, Russia, Saudi Arabia, South Africa, Turkey, the United Kingdom, the United States of America plus the European Union
G77	group of 77 developing countries
GA	General Assembly (United Nations)
GATT	General Agreement on Tariffs and Trade
GDP	Gross Domestic Product
GEF	Global Environmental Facility
GEO	Global Environmental Outlook
GLOBE	Global Legislators Organization for a Balanced Environment
GMEF	Global Ministerial Environmental Forum (UNEP)
GPA	Global Programme of Action
GSP	Global Sustainability Panel
GWP	Global Water Partnership
HDI	Human Development Index
ICC	International Chamber of Commerce
ICE	International Court for the Environment
ICES	International Council for the Exploration of the Sea
ICESDF	Intergovernmental Committee of Experts on Sustainable Development Finance
ICFTU	International Confederation of Free Trade Unions
ICI	Imperial Chemical Industries
ICJ	International Court of Justice
ICLEI	International Council for Local Environmental Initiatives
ICPD	International Conference on Population Development
ICWSD	International Conference on Water and Sustainable Development
ICSU	International Council for Science

IEG	international environmental governance
IFAD	International Fund for Agriculture Development
IFAW	International Fund for Animal Welfare (IFAW)
IFF	Intergovernmental Forum on Forests (United Nations)
IFI	International Financial Institution
IFSD	institutional framework for sustainable development
IGO	Intergovernmental Organization
IISD	International Institute for Sustainable Development
ILO	International Labour Organization
IMF	International Monetary Fund
IMO	International Maritime Organization
INC	Intergovernmental Negotiating Committee
IPBES	Intergovernmental Panel of Biodiversity and Ecosystem Services
IPF	Intergovernmental Panel on Forests
IPCC	Intergovernmental Panel on Climate Change
IPSD	Intergovernmental Panel on Sustainable Development
IRENA	International Renewable Energy Agency
ITFF	Interagency Task Force on Forests
IWC	International Whaling Commission
IWRB	International Waterfowl and Wetlands Research Bureau
IWRM	integrated water resource management
IUCN	International Union for Conservation of Nature
ISSC	International Social Science Council
ITFF	Interagency Task Force on Forests
JCL	Johannesburg Climate Legacy
JPoI	Johannesburg Plan of Implementation
LA21	Local Agenda 21
LDC	least-developed countries
MAFF	(UK) Ministry of Agriculture, Fisheries and Food
MDG	Millennium Development Goal
MEA	multilateral environmental agreement
MOI	Means of Implementation
MRV	Monitoring, Reporting and Verification
MSI	Mauritius Strategy of Implementation
NCSD	National Councils on Sustainable Development
NOAA	National Oceanic and Atmospheric Administration
nrg4SD	Network for Regional Government for Sustainable Development
NWF	National Wildlife Federation
OECD	Organisation for Economic Co-operation and Development
ODA	Overseas Development Assistance
OPEC	Organization of the Petroleum Exporting Countries
OWG	Open Working Group

PIC	prior informed consent
POP	persistent organic pollutant
PrepCom	Preparatory Committee (for UN Conferences and Summits)
PRSP	poverty reduction strategy papers
PRTR	pollutant release and transfer registers
REEP	Renewable Energy and Energy Efficiency Partnership
RIM	Regional Implementation Meeting
SADC	Southern African Development Community
SAICM	Strategic Approach to International Chemicals Management
SARD	Sustainable Agriculture and Rural Development
SCP	sustainable consumption and production
SDG	Sustainable Development Goals
SDPI	Sustainable Development Policy Institute
SEI	Stockholm Environment Institute
SEED	Supporting Entrepreneurs for Sustainable Development
SE4All	Sustainable Energy 4 All
SIDS	Small Island Developing States
SOWS	Southern Ocean Whale Sanctuary
TAI	The Access Initiative (World Resources Institute)
TFWW	The Future We Want
UCLG	United Cities and Local Governments
UN	United Nations
UNCED	United Nations Conference on Environment and Development
UNCHS	United Nations Centre for Human Settlements
UNCIO	United Nations Conference on International Organization
UNCLOS	UN Convention on the Law of the Sea
UNCSD	United Nations Commission on Sustainable Development
UNCSD	United Nations Conference on Sustainable Development
UNCTAD	United Nations Conference on Trade and Development
UNDESA	United Nations Department of Economic and Social Affairs
UNDP	United Nations Development Programme
UNECA	United Nations Economic Commission for Africa
UNSCLAC	United Nations Economic Commission for Latin America and the Caribbean
UNECE	United Nations Economic Council for Europe
UNEO	United Nations Environment Organization
UNEP	United Nations Environment Programme
UNEP	GC United Nations Environment Programme Governing Council
UNESCO	United Nations Educational, Scientific and Cultural Organization
UNFCCC	United Nations Framework Convention on Climate Change
UNGA	United Nations General Assembly

UNGASS	United Nations General Assembly Special Session to Review and Appraise the Implementation of Agenda 21
UNGC	United Nations Global Compact
UN-HABITAT	United Nations Human Settlements Programme
UNIDO	United Nations Industrial Development Organization
UNITAR	United Nations Institute for Training and Research
UNITF	United Nations Inter-agency Task Force
UNISDR	UN International Strategy for Disaster Reduction
UNOSD	UN Office of Sustainable Development
UNWTO	United Nations World Tourism Organization
USAID	United States Agency for International Development
UNU	United Nations University
VC	Voluntary Commitments
WBCSD	World Business Council for Sustainable Development
WCP	World Climate Programme
WDCS	Whale and Dolphin Conservation Society
WEHAB	Water, Energy, Health, Agriculture and Biodiversity
WEO	World Environmental Organization
WFO	World Farmers Organization
WFP	World Food Programme
WHO	World Health Organization
WIPO	World Intellectual Property Organization
WMO	World Metrological Organization
WRI	World Resources Institute
WSSD	World Summit on Sustainable Development
WTO	World Trade Organization
WTTC	World Travel and Tourism Council
WWAP	World Water Assessment Programme
WWF	World Wide Fund for Nature

INTRODUCTION

What makes a hero? On our screens and in books it is often their appearance—their clothes and gear—that give them away. Some slip on shiny, skin-hugging lycra as they take to the skies, cloaks billowing behind them on their way to defeat the supervillain and save the day. Others carry wands or swords, sonic screwdrivers, or laser blasters. Some have turbocharged cars or sleek spaceships.

Real life is different and—in our view—more interesting. Real-life heroes can't call on unearthly powers. They can't win the day with their sword-wielding skills, their sharp shooting, space pilot savvy, or knowledge of advanced alien science.

Real-life heroes are just like you and me. They rely on ordinary, everyday skills—intelligence, integrity, persistence, persuasion, even humour—to achieve the extraordinary. And yet, their task is no less important than their fictional peers. Even without superpowers, many real-life heroes are doing their utmost to save the planet.

And make no mistake, the planet needs saving. Today more than ever, when the world is beset by environmental, social, healthcare, and economic challenges, we need heroes: courageous individuals working both nationally and globally.

This book is about environmental heroes. It tells the stories—some for the first time—of people who have played a global role in taking on an environmental challenge that is larger than their organization or their country. In so doing, they have helped improve our world.

Our work draws inspiration from President John F. Kennedy's book, *Profiles in Courage*. Originally published in 1956, Kennedy tells the stories of eight United States senators who showed tremendous courage at critical moments in America's story, despite the 'risks to their careers, the unpopularity of their courses, [and] the defamation of their characters.'

DOI: 10.4324/9781003202745-1

Updating this for the modern era, our book goes beyond national borders and takes a global perspective. Each chapter tells the story of an individual who has played a significant role in securing a global agreement or landmark treaty on a major environmental threat. We scratch beneath the surface to figure out what kind of person each hero was. What were they like? What motivated them? How did they beat the odds and achieve some sort of success? And what is their legacy today? Interestingly, we uncover a range of characteristics and attributes. No two were the same.

Some of our heroes were humble, while others were larger-than-life figures. Some were excellent listeners, while others were consummate talkers—even raconteurs. Some were ferocious in their single-minded will to succeed, taking no prisoners along the way, while others forged friendships and alliances. All were persistent, refusing to give up even when their situation seemed hopeless. Every one of them recognized that international diplomacy—and persuading the world's governments to take an issue seriously—was the only way to address an environmental challenge that is global in nature. From saving the whales to protecting our ozone layer, from championing a more sustainable way of life to combating climate change, these heroes were determined to make a difference. Each has a unique, fascinating, and at times, even shocking life story.

Heroes of Environmental Diplomacy: Profiles in Courage tackles a different topic— and a different hero—in each chapter. In each story, we review the political and environmental situation the world was facing at the time. We provide insights on each hero's character and motivations, explain how they helped bring the issue to an international audience, and reveal how the situation was resolved. Based on personal interviews and insider stories of those who were there, we show the impact of the individual, assess the state of play today, and explain what needs to happen next.

Why did we write this book? First, as experts who are immersed in the world of international diplomacy, we want to bring these compelling and dramatic stories to a wider audience. Efforts to protect our planet—where they have succeeded—did not just happen. They were *made* to happen by people who cared enough to devote their lives to it.

Second, we want to show the role individuals with courage can play in bringing about *global* change. Each person featured in the book has advanced an international response to a pressing environmental challenge—in the shape of a global treaty or agreement—that has changed the world for the better. Often, they achieved this in spite of daunting hurdles and obstacles. This book presents leaders who, in the words of Ernest Hemmingway, showed 'grace under pressure.'

Third, we hope these stories will serve as an inspiration for the next generation of leaders as our global community seeks to face down the latest—and largest—wave of environmental crises in the 2020s and beyond.

Finally, we want to show that international engagement—diplomacy and negotiations—actually works. Yes, the world is beset with difficulties. But the only way we can deal with global threats is by coming together as a community at a planetary level. Many problems we face today, from climate change to biodiversity loss, are too big for one single country to deal with alone. We can only do this by working together.

While the problems today seem daunting, this book shows we should not give up hope. We have tackled and overcome challenges before. This book tells some of these remarkable stories and explains just how it can be done.

Introducing Our Heroes

Who are the heroes of environmental diplomacy featured in our book? What problems were they trying to tackle? Our book divides up the challenges into four main topics: (i) protecting nature; (ii) human-made chemicals; (iii) climate change; and (iv) sustainable development. Under each of these four areas, we include 2–3 chapters. Within each of the four sections, the book is chronological, with the earliest stories featured first. Here are more details on the sections and the chapters they contain.

Protecting Nature: Geoffrey Matthews, Luc Hoffmann, Eskandar Firouz, and Sidney Holt

In Chapter 1, we introduce an Englishman, an Iranian, and a Swiss who were trying to **preserve our world's wetlands**. Geoffrey Matthews, Luc Hoffmann, and Eskandar Firouz were an unlikely trio who collaborated in the 1960s and early 1970s in a quest to create a new treaty—the Ramsar Convention. Marshes, mangroves, swamps, bogs, and other wetlands don't feature highly on many people's radars. Yet they play a critical role in the health not just of our planet but also humanity, from reducing the impact of floods and coastal erosion to providing 20% of our food. Wetlands are also considered essential for biodiversity, providing homes to a multitude of plant and animal species. How did these three individuals set about protecting these areas? And how did the Cold War—then at its height—threaten their plans? This first chapter sheds light on these unlikely heroes and their legacy, with first-hand accounts and previously unpublished stories of these early days in global conservation.

In chapter 2, we tell the tale of Sidney Holt, a marine scientist who dedicated his 70-year scientific career to **saving whales from extinction**. Holt worked with the International Whaling Commission (IWC) for more than 30 years and is recognized as one of the most influential people in the fight to save these majestic marine species. His story begins in the 1960s when as a member of the IWC committee of 'three wise men' he tried to impose sustainable catch limits. It covers memorable incidents, from the first catch-quotas being applied to Antarctic whaling through to Holt's later work securing the global moratorium on commercial whaling, creating whale sanctuaries in the Indian and Southern Oceans, and ultimately helping end high-seas whaling in 2019. Along the way, we reveal insights on a person who did so much to change humanity's relationship with our planet's great whales.

Chemical Threats: Mostafa Tolba and Franz Perrez

Chapter 3 looks up to the skies and the momentous efforts to **protect our ozone layer**. How did an Egyptian scientist win the support of politicians on both the left and the right to secure one of the most successful and impactful environmental treaties to date? This chapter tells the story of Mostafa Tolba, the mercurial and driven leader who sought to galvanize the world to take seriously the danger posed by chlorofluorocarbons (CFCs)—chemical compounds that pose a major threat to our ozone layer. It shows how he championed the issue in the mid-1970s when scientists first began to suspect CFCs were a problem, through the turbulent 1980s when the scale of the threat had become apparent. How did Tolba persuade cautious governments and obstructive manufacturers, ultimately winning over world leaders like Ronald Reagan and Margaret Thatcher to his cause? This chapter tells the fascinating story of an unlikely victory for the environment—and its legacy and impact in terms of the countless lives it has saved.

In Chapter 4, we look at efforts to **keep humanity safe from hazardous waste**. Indefatigable Swiss lawyer and diplomat, Franz Perrez, has played a major role in this area, as well as several others relating to chemicals management and safety. Through his efforts and those of other leaders in this field, the Basel Convention, which governs the movement of hazardous wastes ranging from disused electronics to old ships, banned their movement and dumping from wealthy countries to poorer ones. The ban, which closed loopholes in the Basel Convention, will have a positive effect on the health and well-being of millions of people, especially in the Global South. How did Perrez and others secure global agreement on this important but complex issue? And what needs to happen next?

Climate Change: Raúl Estrada-Oyuela, Barack Obama, and Christiana Figueres

The next three chapters focus on the biggest challenge of our time: climate change. Each chapter looks back on a different milestone event—starting in 1997 and ending in 2015—and evaluates how these moments moved forward the climate agenda.

In Chapter 5, we look back to 1997 and the role of charismatic Argentinian Raúl Estrada-Oyuela. How did this career diplomat and father of eight come to play a critical role in shepherding through the **Kyoto Protocol**? This chapter looks at how his diplomatic skills–his patience, humour, and tactic of 'negotiation through exhaustion'—yielded results in Kyoto, Japan. It reviews Kyoto's legacy in light of later events, affirming that, while Kyoto did not save the world from the climate threat, it did start us on an important path while creating many of the ideas and approaches that are in widespread use today.

In Chapter 6, we turn to events more than a decade later. In 2009, US President Barack Obama parachuted in at the last minute to a climate meeting that seemed on the verge of collapse. He emerged a day later with a new agreement, the **Copenhagen Accords**. But the agreement was weak and many viewed it as a failure and a missed opportunity. Is this true? Was Copenhagen really the disaster many thought at the time, a moment with more villains than heroes? Or did it sow the seeds for later success thanks to the heroism and courage of Obama and others? Drawing on experts' insights and the stories of those who were in the room, this chapter casts new light on the meeting and examines whether President Obama deserves to be called a hero … or a villain.

Chapter 7 concludes our review of the ongoing climate change negotiations by looking at the last major breakthrough in 2015. The **Paris Climate Agreement** marked the culmination of years of painstaking work and preparations. At the centre of it all was Christiana Figueres, a Costa Rican diplomat who became involved in the climate negotiations in the 1990s, Figueres had taken over as head of the UN's climate office just months after the Copenhagen Summit and had been working tirelessly for five years to craft a more ambitious global commitment to combating climate change. This chapter assesses Figueres' role in building a consensus among the diplomatic community and many other stakeholders, including the corporate sector and non-governmental organizations.

Finally, we close this section of the book by assessing the impact of these three milestones (Kyoto, Copenhagen, and Paris) and evaluating what needs to happen now to build on Paris and on the recent Glasgow Summit in November 2021.

Sustainable Development: Maurice Strong, Maria Luiza Viotti, and Paula Caballero

In our last section, we look at what may be the most ambitious and—ultimately—most important issue of all. Unless we are able to develop sustainably—to end our over-taxing of the planet's resources and address challenges like poverty, inequality, and environmental degradation in a holistic way—then our efforts on ozone, climate change, and other threats will ultimately fall short. Chapters 8–10 look at three events that are bound up inextricably with the sustainability agenda.

In Chapter 8, we tell the inside story of the **Rio Earth Summit**, that momentous event in 1992 when the cause of sustainable development was put firmly on the global table. How did former 'oil man' Maurice Strong come to play such a critical role in the quest to get sustainability on the international community's radar? And what part did he play in helping elevate the voices of many stakeholders—from indigenous peoples to women's groups, youth to unions—in the halls of rich and powerful governments? From Strong's early days growing up in poverty in Canada after the Great Depression, to his rise up the corporate ladder, this chapter sheds light on this unique person and the vital role he played in Rio in 1992.

Leading on from the previous story, Chapter 9 jumps ahead two decades to **Rio+20** in 2012. How did a senior Brazilian diplomat, Maria Luiza Viotti, revitalize the message from Rio all those years before and turn around a UN summit that was in danger of failure? What did Viotti, working with Brazilian colleagues and other allies, do to navigate through a series of potential diplomatic roadblocks and pitfalls, creating a process that would ultimately lead governments to adopt international targets to support sustainable development? This chapter explains Viotti's role, impact, and legacy.

Last but certainly not least, in Chapter 10 we look at the work done both before and after Rio+20 to secure the **Sustainable Development Goals**. The SDGs, as they are widely known, establish 17 new and ambitious sets of goals and 169 targets for achieving sustainable development internationally. From poverty to hunger, health to education, equality to economic growth, and of course, the environment, the SDGs raised the level of ambition for governments and other stakeholders around the world. A major influence on this process was Paula Caballero, a Colombian official now working in the environmental sector in North America. This chapter reveals how Caballero helped advance a complicated and controversial agenda in the teeth of some serious obstacles.

Finally, we end the book by reflecting on our heroes' profiles. What can we learn from them? What characteristics did they all share? What, collectively, did they achieve? And could they have done it all on their own, or were they all really part of a wider circle of friends and allies?

We also reveal who didn't make our final list of heroes. We explain our reasons for selecting these particular heroes and global treaties, and we speculate on whether a similar book written a decade or more from now might feature a different sort of person to those in this version (for instance, more women and more heroes from the Global South). Finally, we end with a look to the future and why we need heroes now more than ever before.

Thank You

We would like to thank the following people for their help in making this book a reality. From Routledge publishers, we are immensely grateful to Hannah Ferguson, Katie Stokes, and John Baddeley for their ongoing support, guidance, and advice. For individual chapters, we warmly thank the following people:

Chapter 1 (Luc Hoffmann, Geoffrey Matthews and Eskandar Firouz): Alison Goddard, André Hoffmann, Andrey Vavilov, Charlotte Touati de Jonge, David Stroud, Eckhart Kuijken, Ed Jennings, Erik Carp, Nick Davidson, and Tobias Salathé.

Chapter 2 (Sidney Holt): Leslie Busby, Vassili Papastavrou, Kieran Mulvaney, Azzedine Downes, Phillip Clapham, Emory Anderson, Roger Payne, and Avery Revere. Author photo credit: Ken Rivard Photography.

Chapter 3 (Mostafa Tolba): Andrey Vavilov, Derek Osborn, Duncan Brack, Keith Ripley, and Nadia Sohier Zaman.

Chapter 4 (Franz Perrez): Felix Dodds, Mariá Christina Cardenas-Fischer, and Franz Perrez.

Chapter 5 (Raúl Estrada-Oyuela): Raúl Estrada-Oyuela, Richard Kinley, Dan Reifsnyder, and Michael Zammit Cutajar.

Chapter 6 (Barack Obama): Ian Fry, Joanna Depledge, Kati Kulovesi, Lisa Schipper, and Richard Kinley.

Chapter 7 (Christiana Figueres): Tom Rivett-Carnac, Daniel Klein, and Conor Barry.

Chapter 8 (Maurice Strong): Patrick O'Hannigan, Hanne Strong, Ashok Khosla, Bill Pace, Derek Osborn, Janos Pasztor, Jeb Brugmann, Julia Marton-Lefèvre, Lesha Witmer, Lloyd Timberlake, Lucien Royer, Michael Dorsey, Michael Strauss, Mirian Vilela, Nitin Desai, Peter Padbury, Tony Simpson, and Yolanda Kakabadse.

Chapter 9 (Paula Caballero): Paula Caballero, Marianne Beisheim, Felix Dodds, Farrukh Khan, and David O'Connor.

Chapter 10 (Maria Luiza Viotti): Amena Martins Yassine, Nitin Desai, Pedro Aurélio Fiorencio Cabral de Andrade, Vicente de Azevedo Araújo Filho, and Paulo J. Chiarelli.

1

LUC HOFFMANN, GEOFFREY MATTHEWS & ESKANDAR FIROUZ

Escaping a Cold War Quagmire: The Ramsar Convention on Wetlands

Chris Spence

Any story that starts with 'an Englishman, an Iranian, and a Swiss walk into a room' sounds like the opening line of a bad joke. In fact, conservationists from these three countries were doing something very serious. They hoped to convince governments to sign a treaty they believed would protect our world.

The year was 1971. The proposed accord was about something most people don't think much about even today: wetlands. These three individuals were determined to persuade the world's governments to take better care of the world's marshes, bogs, swamps, fens, mangroves, floodplains, and peatlands—and the animals and plants that live in them.

Why did a treaty on such a seemingly minor issue matter so much? And how did the Cold War between the USA and USSR—then at its peak—threaten to sink their hopes?

Far from Neutral: An Unlikely Swiss Story

To understand what happened in 1971, we need to go back almost half a century. On 23 January 1923 in the town of Basel, Switzerland, a boy was born. His parents named him Hans Lukas, although in later life he was known simply as Luc. His father, Emanuel, was a successful businessman and his mother, Maja, a talented sculptor. But his family's wealth could not shield Luc from early tragedy. When he was just nine, his father was killed in a car crash. Less than a year later, his older brother died of leukaemia. These two events had a profound impact on the young boy, who found solace in nature and wildlife.

Luc's family owned and ran a pharmaceutical business, Hoffmann-LaRoche. Even then it was doing extremely well. Today, it is a sprawling multinational empire boasting almost 100,000 employees. But in spite of his exposure to this world of big business, young Luc did not seem inclined to follow in his father's footsteps.

DOI: 10.4324/9781003202745-2

After his father's death, Luc's artistic mother remarried the well-known conductor and musician, Paul Sacher. Again, however, their influence did not persuade young Luc to follow them down their chosen path. Already, Luc's own passion was apparent. As a teenager, he would spend countless hours birdwatching around Basel. His interest in nature later found an outlet in academic pursuits. He wrote his first research paper on waders—long-legged shorebirds—at just 18 and went on to complete his Ph.D., with a thesis on the Common Tern, after World War Two.

By then his infatuation had extended beyond waterbirds and into the habitats they inhabited. Through observation and research, Hoffmann had realized the link between his beloved waterfowl and the state of the wetlands they lived in. Hoffmann was not alone. By the 1950s, it was clear many wetlands were in decline. Early warning bells had started ringing in the US in the 1930s as land reclamation, drought, and increased hunting hit geese, duck, and other bird populations. Similar impacts were also being noticed in the UK and across Europe.

As an ornithologist, Luc Hoffmann believed the future of these waterfowl was tied inextricably with the health of the places they lived in. Hoffmann was right. We now know wetlands are valuable to our planet and our own species in many ways, from reducing the impact of floods and coastal erosion to providing 20% of humanity's food, including fish, rice, and many fruits and vegetables. Wetlands are also considered essential for biodiversity, providing homes to a multitude of plant and animal species. In addition, they can play a role in adapting to climate change, by storing carbon that would otherwise be in the atmosphere, and by providing a barrier against flooding and storms.

When Luc Hoffmann's fascination with wetlands began, however, they were hardly beloved places. In fact, they had a pretty bad reputation. These habitats featured negatively in many people's imaginations. In literature, they form an eerie backdrop to Sherlock Holmes' most famous case. In *The Hound of the Baskervilles*, the Grimpen Mire is a fog-shrouded, sinister landscape where animals and even people sink without trace. In *The Lord of the Rings*, the Dead Marshes are a 'forsaken country' of 'sullen waters ... dead grasses and rotten reeds' in which the animated corpses of fallen warriors from an ancient battle lie in wait to drown the unwary traveller.

What's more, there was then little scientific evidence or research to support Luc Hoffmann's belief in wetlands' utility and value. What could he do to change people's minds? With a quiet determination, the young ornithologist set to work.

Hoffmann's first step was to focus on building a research-based case for action. In 1954, he set up his own biological research institute, the Tour du Valat, in the Camargue region of southern France. The institute, which continues its work to this day, conducted extensive research into wetlands and waterfowl. It also helped preserve many species locally, including the greater flamingo.

Next, Hoffmann began widening the coalition for action and connecting with pioneers in the UK, France, and elsewhere who were beginning to gather data on these unloved places. As a scientist, he wanted to encourage greater knowledge and understanding not just in France and Switzerland, but everywhere.

In 1960, Luc launched the next stage in his efforts. Casting his net wider, he arranged a meeting of international experts in wetlands and waterbirds. He called his initiative Project MAR, naming it after the first three letters for wetlands in three languages: French (*MARecages*), Spanish (*MARismas*), and English (*MARshes*). This event provided the catalyst for a series of annual meetings held over the next eight years. Hoffmann encouraged an ever-growing number of experts from around Europe and the world to attend these conferences. His goal: to build an international network dedicated to wetlands.

'He used the Swiss tradition of building a broad base of support,' explained his son, André Hoffmann, when I interviewed him in 2021.[1]

The first conference, which took place in the Camargue in November 1962, was attended by 80 experts from a dozen European countries, as well as from Australia, Canada, Morocco, and the US. The initial goal was to build the knowledge and research base, and more than 60 papers were delivered. But there was more to it than this: Hoffmann wanted to make the outcomes action-oriented, and as a result the conference report included 13 recommendations for next steps. Hoffmann knew the research would be meaningless if it was not applied to the real world. For that, he needed a wide coalition. In fact, Hoffmann wanted to influence *everyone* whose work affected wetlands, from politicians to engineers, farmers, and hunters.

One of the major tools in his armoury was his ability to listen to others.

> When he entered a room, he was there with his characteristic bushy eyebrows, his gentle smile, his natural authority and his outstanding knowledge about wetlands, waterbirds, ecology. Most of the time he was silent, listened, but when he spoke it made sense and advanced the discussion. He was a good listener, a patient teacher, an inspiring mentor and a very practical man,
>
> *recalls Johan Mooij, who worked with Dr. Hoffmann at WWF.*

But Hoffmann had more than the ability to remain silent. He also possessed a steely determination that belied his gentle exterior. While Hoffmann's birth country of Switzerland has a reputation for neutrality, Hoffmann was anything but impartial on the subject of wetland preservation.

'He was always pushing relentlessly forward,' his son André told me during our interview. 'When he presented his case for action, he believed people would either agree or didn't understand … [in which case] he explained it again,' André added.

Luc Hoffmann was also unafraid of making unexpected allies. For instance, he engaged actively with the hunting community. 'This was a contradiction to some scientists, but there was a unity of purpose,' recalls André Hoffmann. This was because hunters wanted wetlands and waterfowl to thrive just as much as the ornithologists did, since hunting needs a healthy environment and abundant wildlife to be sustainable. This type of out-of-the-box thinking and coalition building grew the number of allies to the cause.

The 1962 conference in Camargue was followed by more international meetings, starting in St. Andrews in Scotland in 1963 all the way through to Espoo in 1970. The Camargue conference also led to a ground-breaking document, *Liquid Assets*. Published in 1964, this document outlined the available research and made a strong case for preserving wetlands. It demonstrated convincingly that the world's wetlands play a critical role in both our natural and agricultural systems. Not only do they provide a staggering one-fifth of humanity's food, they are also critical for biodiversity and are widely considered the most biologically diverse of all the world's various ecosystems. They also help protect coastal communities against flooding and shoreline erosion, play a key role in water storage, and have become valued by many people for recreational purposes.

Enter the Russians

Hoffmann's initial coalition of ornithologists was now growing with every conference and each passing year. An increasing number of governments, with the Netherlands and the UK now at the forefront, began to help drive the process forward.

A major breakthrough occurred in 1966 when for the first time a scientist from the Soviet Union officially joined a gathering in Noordwijk, the Netherlands. According to author and researcher Alison Goddard, who wrote an excellent, yet-to-be published account of Luc Hoffmann's life, the Swiss ornithologist was particularly pleased. 'The inclusion of the Soviet Union meant that the major bird breeding grounds in Siberia could be included in the proposed convention,' Goddard observed. It also made it likely that the vast territories of the USSR's allies in eastern Europe might soon join the throng.

It was also a surprise to many given the geopolitics of the era. As Goddard rightly notes, this was at the height of the Cold War and just four years after the Cuban Missile Crisis. The Vietnam War—a proxy conflict for ideological conflict between the US-led capitalist states and their Communist adversaries the USSR and China—was heating up.

Of course, such political issues should not matter to scientists. And yet it is not scientists who sign treaties. It is governments. Could hopes for an international treaty—even one as far distant from politics as an agreement on wetlands and waterfowl—escape the freezing grip of the Cold War?

At first, the involvement of the Russians resulted in further progress. By early 1968, a draft text had been developed by the Dutch and been finessed after feedback from others. Recognizing that a treaty would need to tread carefully when it came to issues of sovereign territory—especially during this time of heightened tension between the two Superpowers—the draft emphasized countries' control over which wetlands they would choose for greater protection. It also backed away from any attempt to sanction or punish countries that did not preserve their wetlands.

In the meantime, Luc Hoffmann was extending his efforts to build the research case for action, using his leadership of the International Wildfowl Research Bureau (IWRB) to launch international counts of bird numbers, which started in 1967. From now on, there would be a bird 'census' around the world each January. This would allow experts to spot trends in bird numbers, thus strengthening the scientific credentials of the movement still further.

What's more, the Russians were now so actively engaged and enthusiastic they even offered to host the next meeting. So far, so good. With such advances on several fronts, Hoffmann hoped for a treaty as soon as 1969.

And that was the moment everything stalled.

A Cold War Freeze: Checked by the Czechs

On 21 August 1968, a plane carrying more than 100 Russian agents landed on the outskirts of Prague in Czechoslovakia. At the same time, tens of thousands of allied troops from Hungary, Poland, and Bulgaria were spilling across the borders, crossing the country rapidly as they sought to occupy Prague and other major centres.

Outnumbered and outflanked, the Czech military offered no resistance. Although popular demonstrations and spontaneous acts of resistance flared over the coming days and weeks, the country was soon under foreign control.

Just what was going on?

The invasion was a response from the Russians and their communist subordinates to Czechoslovakia's efforts to liberalize and democratize. While Czechoslovakia in 1968 was a part of the Soviet Bloc of countries in Eastern Europe, it had recently elected Alexander Dubček to lead its government. Dubček was a reformer and a modernizer. While still avowedly a socialist, he wanted to give more rights and freedoms to his citizens and to move his nation towards a new model of government he called 'democratic socialism.' This movement—dubbed the Prague Spring—caused great unease in Russia, however. Fearful these changes might undermine Russian leadership and socialist solidarity, Soviet leader Leonid Brezhnev ordered his forces to invade and bring the Czech government to heel.

The invasion was heavily criticized, especially in the West. The US and many European governments spoke out against Russia's attack. Even China—formerly a communist ally—was highly critical.

Such matters might seem a million miles away from the preservation of wetlands, but in truth such geopolitical matters often intrude into every area of government.

In this case, the fact that the next meeting was to take place on Russian soil caused immediate problems. Should the meeting still go ahead? Should western nations boycott it?

This was certainly the view of the Dutch Government, which indicated it would not send any official participants to a meeting in Russia.

Within 48 hours of the invasion, Hoffmann had written to participants noting the Dutch withdrawal and suggesting discussions on a draft treaty may have to be delayed. But if the Russians were snubbed, would they, in turn, pull out? The process—which Hoffmann and others had nurtured so painstakingly over so many years, and which appeared tantalizingly close to success—now seemed stalled; blocked by the events in the Soviet bloc.

The Gifted Amateur

While the science and the data Hoffmann and his allies had so carefully and successfully gathered about wetlands was clear on what should happen next, it was equally clear delicate diplomacy would be needed to prevent Cold War geopolitics from sinking all his hopes and dreams.

This was a task for which Hoffmann was not prepared, however. 'The world of international diplomacy was not his thing,' his son André recalls. An excellent listener, Hoffmann had neither the temperament nor the time for the sort of delicate negotiations now required. His plate was already overflowing with a number of other important projects and initiatives. Did he really have the bandwidth to take this on?

Like a sprinter in a relay race, Hoffmann was ready to hand the baton to another. But who could make the running now?

It fell to Hoffmann's friend, Geoffrey Matthews, to take the next lap.

Like Hoffmann, Matthews was born in 1923 and he, too, had developed similar interests in ornithology from an early age. In Matthews' case, however, it was his service during World War Two, rather than a family tragedy, that influenced his future course in life.

Matthews had served with the RAF's Bomber Command during the war, working as a navigator on B-24 Liberators in the Indian and Pacific Oceans. His military career, however, had an unexpected outcome. Faced with the challenge of locating small islands in vast oceans, an unusual thought had occurred to him: if it was this hard for humans with maps and modern instruments to find their way, how could migrating seabirds fly hundreds or even thousands of miles and still find their original nesting spots? Matthews felt he had to get to the bottom of this puzzle. After the war he returned to Cambridge University, graduating in 1950 with a doctorate and staying on to pursue his research into migratory bird navigation.

Appointed in 1955 by Sir Peter Scott to run the Wildfowl and Wetlands Trust in England, Matthews helped advance data gathering of wildfowl populations and made the Trust a key hub for bird research. He participated in Hoffmann's MAR Conferences and by 1968 had become an important participant and trusted accomplice in Hoffmann's efforts to secure an international wetlands treaty. In fact, their relationship was so strong that it was around this time that Matthews was named as the new head of the IWRB—today part of Wetlands International. The IWRB was a group Hoffmann had led for the past decade but felt the need to step back from due to the many other demands on his time.

Neither Hoffmann nor Matthews were trained diplomats, but Matthews, with his wry smile and calm demeanor, did his best to rise to the challenge. A friend of his described his quick wit and great sense of humour—a useful skill for lightening the mood in difficult moments. This is something he would need in the weeks and months ahead.

In his excellent book on the history of the Ramsar treaty, Matthews himself recalls the sense of uncertainty and confusion caused by the invasion of Czechoslovakia. Initially, both the Soviets and the IWRB—which was playing a critical role in organizing the meetings—agreed to postpone the meeting. But then the situation changed. On 14 September, the Russian Steering Committee announced abruptly that the meeting would take place as planned, apparently leaving even some Soviet diplomats wrongfooted. Finally, the meeting went ahead in Leningrad, although with fewer participants, a more regional focus and a less ambitious agenda.

It was what happened next, however, that kept the process from failing and showed Matthews' skills as a diplomat, amateur or not.

Matthews, along with several others from the West, met with a number of Soviet officials over dinner. Together, they smoothed over many of the tensions and agreed that the cause of wetlands protection should not be derailed by the Cold War. In this venture, Matthews found allies on the Russian side just as committed as he was himself, including Russian scientist Yu. A. Isakov.

In fact, it turns out the stress had been so great on the Russian side that one of their key people had suffered a heart attack not long before the Leningrad meeting, complicating decision-making on the Soviet side. Still, they had persisted and an Anglo-Soviet understanding was now reached; both sides would continue to work quietly but persistently towards an international treaty.

'Matthews showed great diplomatic skills,' said Eckhart Kuijken, a Belgian scientist active in the early wetlands negotiations, when I interviewed him in 2021.[2] 'We were not thinking about the Cold War [but]…after 1968 diplomacy was crucial. It was intense work and … Dr. Matthews was one of the heroes who could disentangle the difficult knot,' he said.

Even though Matthews' meeting brought a thaw with the Soviets, these new allies would still need to be patient. 'Things became more difficult after 1968,' Kuijken recalls. 'I personally had difficulties doing my goose research behind the Iron Curtain at that time … even though as a scientist I was not political.'

With new barriers to international travel and genuine differences on the substance of a treaty still remaining, a deal in 1969 or even 1970 now seemed out of the question. Nevertheless, scientists and government officials on both sides of the Iron Curtain continued to work discreetly behind the scenes to develop a text that might be acceptable to all.

A New Complication: The Location Equation

One of the trickiest questions after the Leningrad meeting in 1968 was where a conference to ink a deal might be held. On the surface, this should have been easy. More than a dozen governments were now actively engaged in the discussions. Surely any of them would have been happy to host a conference of this magnitude?

But the Cold War made many regimes sensitive to location. Would the West agree to hold a potentially groundbreaking meeting—an event that might be remembered forever in the name of a treaty—in the Soviet bloc? Equally, would the Russians allow the Dutch, West Germans, Swiss, French, Belgians, or British—all prominent in wetland meetings in the 1960s—garner all the glory? It seemed unlikely. Some out-of-the-box thinking was needed. As an Englishman,

Matthews could not offer to host the event in the UK. Instead, he needed an unlikely ally, someone who would not offend Soviet sensibilities and could defuse this tricky Cold War conundrum.

And this was when our third hero came onto the scene. Just months after the difficulties at Leningrad, Geoffrey Matthews was visited by a scientist and engineer, Eskandar Firouz. Together they hatched a plan they hoped would overcome one of the biggest remaining hurdles to cementing a global treaty. Their idea: that Iran should host the conference.

The Iranian Answer

Eskandar Firouz was born in 1926 in Shiraz, Iran. His family were a part of the Qajar royal dynasty, past rulers of Persia. Firouz himself had been educated overseas, first in Germany and later in the US, where he attended Yale University. An engineer by training, he had developed a passion for conservation and an interest in national parks and protected landscapes.

The suggestion to hold the key meeting in Iran was inspired. Iran at that time was ruled by Mohammad Reza Shah. While the Shah was counted in the pro-Western camp, the fact Iran was neither a Western country nor a member of any major anti-Soviet coalition such as NATO made it more palatable to everyone. Iran may not have been exactly neutral, but neither was it central to any Cold War calculus. When Matthews and Firouz floated the idea to other delegates, it was generally well received.

Firouz's own character and demeanor helped considerably. Like Matthews, Firouz possessed a talent for finding common ground and compromises people could accept. One contemporary I contacted recalled Firouz's great authority and expertise. Combined with his pleasant and friendly manner, it produced the perfect combination of skills prized by any effective negotiator.

Firouz would need these skills in abundance once the meeting began. As scientists and diplomats from around the world gathered at the pleasant coastal resort of Ramsar on the Caspian Sea in late January 1971, there were still several difficult and thorny issues to resolve.

For a start, different draft texts offered by the Dutch and the Russians needed to brought together. As the conference Chairman, Firouz did a marvellous job navigating these differences and helping delegates find agreement. 'He was so clever and technically knowledgeable and knew how to present the ideas of others effectively. He could give brilliant summaries of their ideas so they were acceptable to others,' recalls Kuijken, who holds the distinction of being the youngest official present at the meeting. Firouz also led by example, announcing during the meeting that Iran had earmarked an important wetland area of 130,000 hectares for protection.

One dispute was over how much of a focus the treaty should place on protecting birds. Many environmentalists had moved beyond the idea a future treaty should concentrate mostly on the migratory birds that use wetlands. While birdlife might have been their starting point 20 years before, by 1971 most experts now saw the wider benefits wetlands protection and their wise use could bring. Their vision has since been borne out—wetlands are now viewed as crucial for everything from food production to reducing coastal erosion, combating climate change, and their wider ecological and social value. Clearly, these impacts go far beyond that initial framing of a hospitable place for migrating birds. But in 1971 the USSR, in particular, wanted a strong emphasis in the treaty on waterfowl.

Why? The Russians had some brilliant scientists and some, like Yu. A. Isakov, were just as visionary and progressive as their Western contemporaries. 'Isakov was inspiring and a good colleague with extraordinary knowledge,' Kuijken confirms. So why did the Russians wish to highlight the role of waterfowl?

During the course of my research, I came across two theories. Both suggest Russia's internal politics may have been at play.

In one interview, an expert who had attended some of these meetings told me he had heard that if waterbirds were explicitly mentioned in the treaty, it meant a particular Soviet ministry would retain authority over the treaty's implementation; if not, they would lose control to a different, perhaps less conservation-minded, department.

Meanwhile, in another version of events, the Soviet scientists involved felt they could get the treaty rubber-stamped back in Moscow if it included a stronger reference to waterfowl because they could then say to their political bosses that 'it's just about birds.' Without the bird references, it would be a treaty about land—which would trigger a far higher level of political interest and could scupper the whole deal.

Was the inclusion of stronger references to birds a final ploy by passionate Soviet scientists to keep the Cold War and interfering politicians from getting in the way of a treaty they knew would benefit the planet? Whatever the real reason, the Russians got their way. On 2 February 1971, the *Ramsar Convention on Wetlands of International Importance especially as Waterfowl Habitat* was signed by delegates from 18 countries. While waterfowl are still included in the official title of the treaty to this day, we now know beyond any doubt that the treaty is much more about land management than birds, important though ducks, swans, and our other avian friends may be.

'A Victory for Science'

The day itself felt momentous for those present.

'I remember the signing ceremony in Ramsar—my boss actually signed it first since Belgium was ahead of other countries alphabetically,' Kuijken recalls. 'And I clearly remember brilliant people like Matthews, Hoffmann, Firouz, and Erik Carp.[3] As delegates lined up to sign the document, I recall feeling such a sense of belief and relief,' Kuijken said. 'Belief we could achieve such things for conservation and relief that we had finally done it. It was a victory for science over politics.'

At the heart of the treaty from the very beginning was the concept of 'wise use' of these important places, including the conservation and sustainable use of wetlands, the services they provide for people and their importance in our natural ecosystem.

The treaty—one of the first global agreements to conserve and protect nature—had a modest beginning. On that fateful day in 1971, it had signatories from just 18 countries. But by 1975 it had become a legally recognized international treaty, by 1980 the number of countries to join the treaty had risen to 28, and by the turn of the millennium it had 114. Today, more than 170 countries have ratified the treaty and are actively protecting wetlands.

While countries are not compelled to protect all of their wetland areas, progress over the past 50 years has been impressive. Because of that early work by Hoffmann, Matthews, Firouz, Carp, Isakov, and other pioneers, today the Ramsar treaty helps safeguard more than 2,400 individual wetland areas covering a combined area of 2.5 million square km. To put that into perspective, that's bigger than Mexico, twice the size of South Africa, and the same size as 19 Englands.

'The impact of Ramsar was huge in so many parts of the world, from Senegal to Costa Rica and so many other places. It was really a turning point,' says Kuijken.

André Hoffmann agrees: 'My father was proud of what it achieved,' he affirmed (Figure 1.1).

FIGURE 1.1 A moment in history—the Ramsar Signing Ceremony, February 1971. Left to right: V. D. Denisov (USSR, Vice-President), Eskandar Firouz (Iran, President), F. Liebenberg (South Africa), M.F. Mörzer Bruijns (the Netherlands, Vice-President), G.V.T. Matthews (UK, Rapporteur General) and Erik Carp (IWRB, Administrator). Photograph by Eckhart Kuijken

A Heroic Legacy

And what of our three heroes? How did each of them fare after the Ramsar treaty was signed on that fateful day in 1971? Their futures took quite different paths.

Having already achieved so much on the international stage—not only with Ramsar but in helping found the IWRB and then WWF in the 1960s—Luc Hoffmann continued for many years to have an influence globally, especially in Europe and Africa. In 1994, he established the MAVA Foundation, which supports nature conservation across the Mediterranean as well as in the European Alps and on Africa's west coast. Meanwhile, he continued to write prolifically, authoring more than 60 books. In 2012, the Luc Hoffmann Institute was established to act as a catalyst for innovative efforts to support biodiversity around the globe. By the time of his passing in 2016 at the age of 93, he had received accolades and awards worldwide, including honorary doctorates and the French Legion of Honour, that country's highest order of merit.

Geoffrey Matthews also continued to have a major influence, playing a role in the creation of Europe's 1979 Wild Birds Directive, which drew inspiration from the Ramsar treaty, as well as the UK's 1981 Wildlife and Countryside Act. He remained in leadership roles at the IWRB and the Wildfowl and Wetlands Trust in England until 1988, ensuring science and research was front-and-centre in government decision making. He also helped protect species such as the Greenland White-fronted Goose and Eurasian Curlew. As if this wasn't enough, he also wrote over 150 papers and chapters in 43 books, and in retirement he penned *The Ramsar Convention on Wetlands: its History and Development*, which remains the most informative and authoritative account of the treaty yet written. In 1986 he was awarded an OBE by Queen Elizabeth. In a demonstration of how respected he was by the conservation community, he even had a previously unknown bird louse, 'Onithobius matthewsi,' named after him. He passed away in 2013 at the age of 89.

For our Iranian hero Eskandar Firouz, the future after 1971 was both more surprising and more hazardous. Initially he enjoyed remarkable success, playing a key role in the creation of Iran's Department of Environment, which he led for many years. During his tenure, Iran developed its Environmental Protection Law, which is still operating to this day. He also played a role in creating Iran's national park system, nature reserves, and other protected areas. Internationally, he was vice president at the famous UN Conference on the Human Environment in Stockholm in 1972, as well as a leading figure at the International Union for the Conservation of Nature (IUCN).

But the 1979 revolution in Iran changed everything. The Pahlavi dynasty was overthrown and replaced by an Islamic republic under Ayatollah Ruhollah Khomeini. Given his family connections to the former dynasty, Firouz was seriously affected and lost his role in Iran's government. Worse was to follow when he was incarcerated by the new regime. Refusing to let his circumstances deter him, he continued his vocation even in prison, where he reportedly taught foreign languages (he spoke several) to his fellow inmates, while also instructing them on the importance of conservation. After his release from jail, he was placed under house arrest. Eventually, he returned to the US, where he continued his conservation work and wrote several books. He died at a hospital in Maryland with his family around him in March 2020 at 93 years old; the same age as Luc Hoffmann.

With our three heroes all living into their late 80s and early 90s, one wonders whether ornithology and conservation might be a good career choice for those seeking greater longevity? Whether this is the case or not, Hoffmann, Matthews, and Firouz would all have been able to look back in pride at what they achieved in the Iranian resort of Ramsar half a lifetime ago.

The Future of Wetlands

And what of the treaty itself? In spite of the progress made during the past half century, the work of the Ramsar Convention is not yet done. In fact, the path ahead looks steep. While more than 2.5 million km^2 of wetlands have been listed under the Ramsar treaty, as many as 10 million km^2 are still unprotected. Many wetlands are under pressure from the spread of cities, as well as from pollution and runoff from farms and factories, and new invasive species migrating as our planet heats up. Climate change is magnifying many of the challenges in wetlands preservation since it is already having significant and unpredictable impacts on regional water systems and bringing more extreme weather. While the decline in wetlands has been slowed in some parts of the world in recent years, thanks in many cases to the Ramsar treaty, many wetlands continue to suffer due to humans and our actions. Research suggests more than one-third of the world's wetlands were lost between 1970 and 2015.

In addition, the fact that a particular wetland may have been listed by a country for protection under the Ramsar Convention does not always mean it is fully safe. In most countries, the Ramsar treaty is considered a 'soft' law, meaning it is often difficult to enforce its provisions in court and is not as binding or regulated as countries' traditional 'hard' laws. This may need to change if wetlands are to be protected more vigorously. Clearly, much remains to be done.

Still, there remains room for optimism. 'My father would want to see wetlands more than ever be part of the solution, especially after the pandemic,' says André Hoffmann. 'Humanity needs a functioning life support system and he would be excited at the opportunities to show the value of wetlands.'

There is no doubt the next decade will be a critical time for the planet, particularly as we strive to address climate change and its causes. To be successful, efforts to conserve and steward our wetlands will need to be taken to a whole new level—something our three early heroes of the Ramsar treaty would have wholeheartedly endorsed.

Notes

1 Interview with André Hoffmann via video conference, 22 June, 2021.
2 Interview with Eckhart Kuijken via video conference, 18 June, 2021.
3 Erik Carp, who was the Administrator at IWRB from 1969, played an important role before, during, and after the Ramsar conference, working closely with Geoffrey Matthews, Luc Hoffmann, and others. He acted as secretary at the Ramsar meeting and was also the editor of its proceedings.

Background References

Boere, G., Galbraith, C., and Stroud, D. (editors) (2006) Waterbirds around the World, The Stationery Office.

Breiding, J. (2016) Luc Hoffmann, Unsung Hero of Nature Conservation, The Ecologist. Available online at: https://theecologist.org/2016/nov/23/heavens-eyes-luc-hoffmann-unsung-hero-nature-conservation

Goddard, A. (2013) The Unsung Ornithologist: Luc Hoffmann and the Role of Big Business and Science in the Birth of Environmentalism, pending publication.

Kayhan Life. (2020) Obituary: Eskandar Firouz, Founder of Iran's Department of Environment, Dies at 93, Kayhan Life News Service. Available online at: https://kayhanlife.com/obituary/eskandar-firouz-founder-of-irans-department-of-environment-dies-at-93/

Kuijken, E. (2006) A Short History of Waterbird Conservation, Ramsar. Originally a chapter in Waterbirds around the World (see Boere. G, see above). Available online at: https://www.ramsar.org/sites/default/files/documents/library/wurc_kuijken_history.pdf

Matthews, G.V.T. (1993) The Ramsar Convention on Wetlands: Its History and Development, Ramsar.

Mooij, J.H. (2016) Obituary: Luc Hoffmann, The Goose Bulletin, Issue 21, December 2016.

Rees, E., and Smart, M. (2013) Obituary: Geoffrey Matthews, OBE, Wiley Online Library. Available online at: https://doi.org/10.1111/ibi.12068

2

SIDNEY HOLT

The Long-Haul Hero Who Saved the Great Whales: The International Whaling Commission

Patrick Ramage

"There is a problem, Signora, on the account of your client – Holt."

"Ah, si," replied Antonella Papini, the bank manager, *"and the problem?"*

"Books," said the man from account security. *"Books ordered from Amazon; dozens of them. 3, 5, 10 books every month. All different titles, all charged on his account."*

"And the problem?" Papini repeated.

"Let me tell you, Signora, there has clearly been fraudulent activity on this account, a man of 90 years cannot read this many books."

"Let me tell you something," snapped Papini, indignant, "You don't know my client!"[1]

On a sweltering July afternoon in 2016, 33 kilometres to the south, the client in question sat moored to an old wooden desk, a life raft of sorts, afloat on a sea of books. There were hundreds of them; bursting from his bookshelves, flowing over the furniture, cascading down the stair-steps from his bedroom and flooding the floor. An Amazon River of non-fiction – carrying new hardbacks and paperbacks, old pamphlets and heavy tomes to every corner, cranny, nook and roof tile of his small stone farmhouse – best not to mention the four rows of impossibly full double bookcases tucked in the shed – towards the bottom of an unpaved path just above the olive grove, steps down the hill from the medieval Umbrian village of Paciano.

Through failing eyes and thick prescription lenses, the farsighted father of fisheries science and saviour of the great whales squinted at a leviathan computer screen. Deploying ancient fingers across plastic keys, patiently hunting and harpooning each letter, he typed:

DOI: 10.4324/9781003202745-3

"Dear Jonathan Balcombe,

My attention was recently drawn to some papers of yours in the on-line journal Animal Sentience and a new book, which I have ordered.

I am a 90-year-old British fisheries biologist living in Italy. A sort of hobby in recent years has been self-awareness in species other than primates, elephants and cetaceans. I was interested, too, in the question of whether fishes feel pain (The obvious cruelty of it led me to spend many years fighting to end whaling). Years ago I decided to oppose all fishing for 'sport' but now seriously wonder whether I should be easing out of the matter of managing commercial fishing. Right now, I'm inclined to think it best to work for very 'conservative' fishing for several reasons, as I worked for conservative whaling in the years when there seemed to be little chance of ending it.

So, I shall follow your work and hope to correspond from time to time – though I of course do not know how long I have got!

Whatever, best wishes. Sidney Holt"[2]

Sidney Joseph Holt made our species' natal transition from marine to terrestrial mammal and drew his first breath on Sunday the 28th of February 1926 in the East End of London. By the time he arrived to save them, the plundering of our planet's great whale populations was well underway.

Sidney's courage defies reduction. It cannot be captured in a single act, decision, or moment. It was made manifest in a lifetime of contributions to marine conservation so massive they render others Lilliputian: developing and publishing the foundational understanding of fish population dynamics (1957), first devising, then tirelessly driving dramatic reductions in international catch limits for commercial whaling (1960 onwards), conceiving, drafting and guiding creation of the Indian Ocean Whale Sanctuary (1979), drafting and midwifing passage of the global Moratorium on commercial whaling (1982), drafting and championing creation of the Southern Ocean Whale Sanctuary (1994), rallying scientist and International Whaling Commission (IWC) member support in creation of the IWC Conservation Committee (2003), advising Australian and New Zealand officials involved in the successful prosecution of their case before the International Court of Justice (ICJ) resulting in a judgment ending Japan's scientific whaling program in the Antarctic.

These monumental marine conservation milestones would be fully consummated (and celebrated) on the 26th of December 2018 – in the 92nd year of Sidney's long and happy life – when the Government of Japan announced its complete withdrawal from high seas whaling, finally ending the commercial slaughter of whales in the Southern Hemisphere and international waters worldwide.

Cometh the Hour, Cometh the Man

Coastal communities around the ocean planet had been conducting shore-based subsistence whaling for millennia before 11th century Basque sailors began pursuing slow-swimming North Atlantic right whales to trade in their oil and parts. These first commercial whalers would be joined first by the Dutch, then the British, then the Americans, then the Norwegians and others.

By the 1840s, sperm whale oil lit the lamps of the western world, and some 900 whale ships were engaged in the grisly industry. Of these, more than 700 were American vessels, voyaging across the global ocean, sacrificing the comfort, safety, sometimes even the sanity of their crew to keep investors in their nation's fifth largest industry awash in wealth.

Unbridled commercial slaughter fed voracious American and European appetite for whale oil and baleen products – from venetian blinds to buggy whips, corsets, scrimshaws and collar stays – reducing the right whale population from an estimated population of 200,000 animals to the very edge of extinction. When the Holt family celebrated little Sidney's first birthday, fewer than 100 northern right whales remained. Gray whale populations in the eastern North Pacific and Arctic Ocean were also severely reduced.

Humpback and sperm whales were next in the crosshairs. Time and again over the following decades, as abundant species became scarce, industrial whalers shifted their gunsights, adding the next great whale species to the target list. Technology transformed the lucrative industry over the ensuing period, often driven by Norwegian innovation. Steam-powered ships enabled hunters to close in on the faster blue, fin, humpback and sei whales while new explosive harpoons added range and accuracy.

At the dawn of the 20th century, having laid waste to numerous whale populations in the northern hemisphere, the whaling ships headed south. Venturing into the biologically rich waters of the Southern Ocean, they discovered unprecedented concentrations of great whales feeding in the Sub-Antarctic and Antarctic zones of the South Atlantic during the astral summer.[3] As Drs. Phil Clapham and Yulia Ivaschenko would later write: "modern whaling had found its great playground, and a slaughter almost unparalleled in the history of wildlife exploitation was about to begin."[4]

The violent slaughter in the Southern Ocean intensified human violence elsewhere. Soaring demand for glycerine-based explosives during the First World War was fulfilled with whale oil from British and Norwegian Antarctic whaling.

In the northern summer of 1925, accelerating innovation again turbocharged the industry. The *Lancing*, a Norwegian factory ship, was fitted with a stern slipway enabling pelagic whaling operations – the hauling up and on-board processing of dead whales on the high seas – obviating the need for whalers to tow their carcasses to a land station or anchored factory ship. The bonanza was on.

Glimmers of Global Conservation

"As commercial whaling ballooned in size and extent," Sidney Holt would later note in his memoirs, "the first stirrings of concern about its impacts began to be heard" (Holt, 2016).

Even before pelagic operations had begun, scientists and some national del-egations on the Whaling Committee of the International Council for the Ex-ploration of the Sea (ICES) – the only international body following whaling at all – had begun to express their worries.

With nudging from ICES and through a process of its own involving the strengthening and expansion of international law in the post-World War I world, the League of Nations also took up the matter.

In 1924, the League established what was to be called the Committee of Experts on the Progressive Codification of International Law (CEPCIL). The committee was charged *inter alia* with preparing a provisional list of subjects:

> the regulation of which by international agreement appeared most desir-able and realizable [and] to submit a report to the Council of the League on questions which appeared sufficiently ripe for solution by conferences.[5]

Among the experts appointed to CEPCIL was Professor Juan Leon Suarez, a dis-tinguished Argentinian jurist and Dean of Political Sciences at the University of Buenos Aires. It fell to Suarez to respond to the question of whether it is possible to establish by way of international agreement rules regarding exploitation of the products of the sea.

Suarez filed his report in December of 1925 – the same month the *Lancing* caught its first whale in the Antarctic and two months before Sidney Holt – the hero who would tighten the noose on high-seas whaling – was born (Holt, 2016).

Throughout his memoirs and other writings, Sidney's sense of kinship with Suarez, whose torch he would later carry, shines through:

"He criticized the centuries old doctrine of 'freedom of the seas', then still the prevailing rule, as the basis for the management of living marine resources, as-serting 'the riches of the sea' and especially the immense wealth of the Antarctic region are the patrimony of the whole human race," and that harmonised global rules were needed because "animals, happier in this than men, are ignorant of jurisdiction and national frontiers and observe not international law but interna-tionalism; the sea for them is a single realm."[5]

Suarez called for coordinated international regulation, stressing the need to bring biological information to bear in decision-making, not just political and economic considerations.

With regard to whaling – which he considered a "barbarous manner of kill-ing" – Suarez made a number of specific recommendations, calling for an agree-ment to settle "such important matters as the protection of young whales, the creation of reserves for adults, and the elimination of waste, especially through the full utilization of all parts of the captured whales."

The Committee of Experts' final recommendations to the League's Council and an ensuing 1927 resolution instructed the Economic Committee of the League to study, in collaboration with ICES, whether and in what terms, for what species, and in what areas international protection for marine fauna could be established.

In that consultation it was decided that "only in regard to whaling was there a clear sense that the action of a multilateral nature and applied on a global basis was needed." ICES "saw a situation different from that of fishing since it was clear that the migratory range of whales was so wide that desirable regulation would probably need to be adopted universally in order to be wholly effective."

This early awareness of the special biological and behavioural characteristics of whales – including, for most species their vast migrations – dictated the need for particular conditions to be applied to the management of their exploitation and conservation, would be echoed decades later in the special provisions accorded to cetaceans as highly migratory species under the United Nations Convention on the Law of the Sea (UNCLOS), in Articles 65 and 120.

The Economic Committee, in association with ICES, undertook to draft a convention for the regulation of whaling. This was done in the course of an experts' meeting in Berlin in 1930. The final version of the new convention was adopted by the League Assembly in 1931 and opened for signature in January 1932. However, it wasn't until 1935 that it came into force, by which time it was clearly insufficient. The long delay was caused primarily by the UK, which did not ratify the agreement until late in 1934; like Japan (which never signed), the UK had continued to argue that the new rules would best be established by bilateral and multilateral conventions among the nations most concerned. In fact, that is what happened in the absence of an international agreement (Holt, 2016).

Early Life and Education

Born a cockney, Sidney was the only child of Sidney and Ethel Maude Holt, respectable working class East Enders who dreamed of a better, happier life for their son. When Sidney was four, the family moved to a new suburb in northwest London where he grew up. His father was half-employed through the 1930s. "I knew what it was like to go down to the labor exchange and collect the dole and that sort of thing." Sidney would say later, "We weren't hungry; we were reasonably comfortable. But we didn't have many books in the house or anything."[3]

Book readers or not, Sidney's parents were bound and determined their son would be well educated. The boy was good at taking tests. When the time came for his 11-plus exams, when English children were "streamed" into different schools based on their results, a teacher encouraged him to sit the exams for Haberdasher's Boys School, where he received a scholarship and began his secondary education in 1937.

By 1940, Sidney, the school's top sprinter and an aggressive rugby player, had become what he later called "adopted middle class." "I began to distance myself from my parents. I still loved them, but I began to be different from them."[3] In his third year, war came, bringing blackouts, food rationing and flashes of anti-aircraft fire in the night sky.

Reduced curriculum during wartime required choices, one of which, in a near miss, set Sidney on a trajectory to his future calling. Reporting one morning the boys were asked to stand in one of two lines indicating their choice of study: history or geography. Sidney chose history. Two more lines would be formed, one for boys choosing science, the other for woodworking.

Sidney Holt, future phenomenon of fisheries biology, co-author in his early 20s of one of the most widely cited scientific texts ever published, eventual author of more than 400 scientific papers, book chapters and popular articles in the fields of fisheries science and management, conservation, protection of marine mammals chose … woodworking.

His name is not familiar, but generations of fisheries and marine scientists, diplomats, managers and advocates can all be deeply grateful to Sidney's best friend and schoolmate Phil Thomas "another scholarship boy, from Wales," for walking to the other side of the room and standing in the science line. At the very last instant, Sidney dashed to join him. "It was only because he was there," Sidney confessed, 70 years on, "I was much more interested in woodwork really. I was building and constructing things all the time, but science as such – I didn't really know what that was."[3]

During his final year at Haberdashers, Sidney started his migration towards biology, reading about evolution and works by popular scientists. On graduation he was accepted to the redbrick University of Reading, where he continued to be immersed in political activities, campaigning for Communist and Labour candidates during the post-war election.

As his voracious reading of history, science and politics continued, Sidney was deeply influenced by *The Social Function of Science* (1939) by J.D. Bernal, a prominent scientist, public intellectual and communist activist. Bernal argued that science could no longer be just a protected area of abstract intellectual inquiry. Science would instead be the chief change agent in society with an inherent function: the improvement of life for mankind everywhere.

Bernal's words resonated deeply in young Sidney; his left-wing sensibilities turbocharged by a generational wave of post-war optimism. "Things were happening, people coming out of the armed forces and certainly those of us at university were committed to making a newer, better world."

That bold vision and fresh thinking drove a lifetime of work. Contrary to later assertions by his critics, Sidney Holt did not convert from the priesthood of science to impassioned advocacy. He was an advocate and change agent from the outset whose scientific work had a political dimension right from the earliest days. Like none before him and too few since, Sidney Holt seamlessly blended public advocacy with peerless science, creating a powerful propellant for change.

During his career of work with the UN Food and Agriculture Organization (FAO), the United Nations Environment Programme (UNEP) and throughout his lengthy engagement with International Fund for Animal Welfare (IFAW) and other leading NGOs, Sidney's relentless refrain was that progress and policy change required a mix of rigorous science, sophisticated politics and campaigns that galvanised public support. As environmental journalist Kieran Mulvaney, a former Greenpeace campaigner and founder of the Whale and Dolphin Conservation Society (WDCS) would later note:

> Sidney was among the first accomplished scientists to show that being objective did not equate to being neutral. He always followed the science but was not afraid to shout from the rooftops about what that science was saying, even if that made some uncomfortable. He provided the intellectual space for the likes of Greenpeace, WWF, IFAW and others to campaign for an end to commercial whaling; he nurtured and encouraged them but was never shy to castigate or correct them if he felt their positions were straying away from the defensible. That could sometimes earn him a reputation for being contrarian, but the truth is that all along, he was being a scientist.

IFAW President Azzedine Downes, an eager student and regular correspondent with Sidney through his later years, agrees.

"By the time I met him, his scientific credentials were beyond reproach, so he had abandoned, if it was ever there, a reticence to appear an advocate. His science was unassailable, and his activism shone through."

Sidney accomplished his first degree at 17 years of age with concentrations in Biology, Zoology and Chemistry. Receiving his second degree (in Zoology) a year later, he hoped to continue his studies and become a veterinary surgeon, which would require four more years of study.

His parents could not afford to support him for such a period, so Sidney interviewed and was recruited to join the UK Ministry of Agriculture, Fisheries and Food (MAFF) Fisheries Directorate at Lowestoft in East Anglia. Veterinary science's loss would be marine science's leviathan gain.

From Three Lumps of Coal to Committee of Three

It was during Sidney's years with MAFF that he produced the most influential book in Fisheries Science, *On the Dynamics of Exploited Fish Populations*, in collaboration with fellow MAFF scientist, Raymond J.H. Beverton.

"Neither Ray nor I knew enough maths to solve the equations we were to work with. Ray went off to Cambridge for a time to complete his studies. So, I was left alone with a book called 'Teach Yourself Calculus' which I'd bought at a local bookshop. That got me through integral calculus."

"When Ray returned, we started working away. It turned out we were a very good pair together. I hated writing, I just wanted to get on with solving puzzles. Ray was a very good writer. The result was, I think, a marvelous piece of writing. He did the writing, not me. I did most of the calculations and things in it. It was a terrific partnership."[3]

"Three lumps of coal was the fuel brought daily into our little office in an annex to the main lab to keep us warm for the day's work through the winter of 1947/48," Sidney would later recall.

"We often worked in overcoats, wearing woolly hats and gloves converted to mittens with the finger-tips chopped off. Wielding a pencil was a special art."[6]

So too the book, which quickly became the most quoted paper in the history of fisheries science and a must-have textbook for fisheries scientists. This Genesis of the modern age-structured approach to the optimal management of fishery resources led directly to the formulation of a fishery catch equation with almost universal applicability, pointing the way to sustainable fishing 50 years before it became fashionable.

The impact of Sidney's first masterwork remained powerful those 50 years later as marine biologist Vassili Papastavrou, his closest collaborator during Sidney's decades of work with IFAW would poignantly recall:

> I was in Brussels with Sidney going round various offices in the EU. We were to meet someone in the Directorate General for Maritime Affairs and Fisheries to discuss what the EU might do about whaling. The meeting had taken some time to organize and we were well prepared. But the person we were meeting was distracted and did not seem to be listening at all. We were not getting through. After a while he asked, "are you THE Sidney Holt?" He turned around and reached from his bookshelves for Sidney's book for Sidney to sign for him. He couldn't really believe that he had Sidney Holt in his office. It was the only fisheries book he had taken with him when he moved to become a bureaucrat in Brussels. Once that was out of the way, he was able to concentrate on what we had to say.[7]

International Whalers' Club?

From its very founding in 1946, the International Convention for the Regulation of Whaling (ICRW) – one of the very first post-war international conventions to do so – enshrined the concept of whales as an "inheritance" of future human generations, as being of common concern to all nations and not just those few involved in their exploitation. Its preamble is unambiguous: "recognizing the interest of the nations of the world in safeguarding for future generations the great natural resources represented by the whale stocks."[5]

In a dinner speech to the negotiators of that Convention, convened in Washington, DC in November 1946, the Assistant Secretary of the US Department of the Interior, C. Giraud Davidson offered this upbeat summary of the task that lay before them:

"Conservation is one of the truest symbols of mankind's hope. No one without conviction of a peaceful and more happy future for mankind would spend time and thought and energy in trying to provide for future generations."[5] Meanwhile, as the delegates finished their desserts and coffee, the rapacious slaughter on the high seas continued.

Enter the Hero

Sidney first entered the political whirlpool of the IWC in 1960 when he was selected one of the Committee of Three – Cof3, the so-called "Three Wise Men" – a committee of independent scientists tasked by the IWC to undertake an assessment of the whale stocks and devise sustainable limits for Antarctic whaling.

The resolution establishing the Cof3 stated: "It is the intention of the Commission in setting up this special group of scientists that the Commission should, not later than 31st July 1964, bring the Antarctic catch limit into line with the scientific findings…"

The urgency of the task was stark. "Our first analyses were so distressing," Sidney later wrote,

> that we decided to submit, early in the 1960s, an urgent interim report saying it would be necessary to reduce the limit drastically and immediately. That was "noted" at the Commission's 1962 meeting in London, but we were unable to report more because the necessary funds had still not been paid so we could not meet.[5]

In a second Interim Report the Cof3 wrote that "Our general conclusions… point to the need for action so drastic and of such urgency that we thought it essential to give the Commission ample opportunity to consider the implications including what action would be needed *before* 1964 (emphasis added)."

The whaling countries, including Japan, the USSR, Norway, the UK, the United States and Australia continued to delay meaningful action, pleading that sharp reductions in catch limits would have catastrophic economic implications given major investments in vessels still to be recouped. Eager for a reprieve, they pressed for more analysis.

Sidney was having none of it.

Addressing the plenary at the 1964 IWC meeting in the Norwegian whaling capital of Sandefjord he laid out the stakes:

> Our best estimates of what is happening to a stock come when the stock has virtually disappeared, and we have a nice long series of data showing its decline. If you wait until you have better estimates next year and better estimates the year after that you will only make decisions when it is much too late, certainly too late to act painlessly from the point of view of the whaling industry. Then, in any case, a new factor comes into consideration as it has with the blue whale, that you risk total extinction.[5]

Despite his exhortation no catch limits were adopted. At the end of the meeting, Sidney, this time speaking on behalf of FAO had the last word. In his statement, he highlighted the stark difference of views on the legal status of the great whales.

> Do they, in effect, belong to no one or do they belong to everyone? Do they, in effect, belong to future generations or to this generation? Do they,

in effect, belong to those who exploit them or to all nations who in some way, however small or indirectly, depend on products of the sea for the nourishment of their people?

These problems seem to us to be inherent in the present status of International Law and our fear is that these germs of their destruction might be in other treaties that exist for conserving and managing the fisheries resources of the high seas. We put this question then: Will the failure of this Commission to agree this year on an effective measure in the Antarctic nourish these germs, or will it set in motion a process of reconsideration by nations of the basic principles of our treaties?[5]

"A Commando Raid": Creating the Indian Ocean Sanctuary

"Small" countries can achieve massive things in international conventions. Like the independent experts and NGOs they sometimes rely on, smaller countries are more nimble and can move more quickly than the more ponderous pace of the great powers. Such was the case in 1978 when Sidney collaborated with the newly independent Government of the Republic of Seychelles on a gambit to secure sanctuary for whales in the waters of the Indian Ocean.

Thanks to recent changes in the Law of the Sea convention, Seychelles had acquired a huge Exclusive Economic Zone (EEZ). The government promptly decided to accede to all relevant international agreements concerning the ocean, including the IWC, to which it adhered on the 19th of March 1979 just four months before it would attend its first IWC meeting in London. The country had a strong conservation record, including the creation of marine parks and reserves and specific protective legislation for its birds, turtles and giant tortoises.

Together with renowned South African conservationist Lyall Watson, Sidney flew from Zurich to the Seychelles capital, Victoria, on the main island of Mahe for a meeting with President France-Albert Rene on what initiative his government might take at its first IWC meeting four months hence.

> We knew the IWC Scientific Committee would meet in Cambridge just before then and that it would be necessary for any initiative with scientific aspects to clear that hurdle. Evidently, successful action would need to be more or less like a commando raid.
>
> *(Holt 2016)*

During the flight, the two discussed ideas they might offer the President. Given the status of whale stocks in the region, they proposed an Eastern Indian Ocean Sanctuary (not a great fit for a Seychelles initiative from the African side). Sidney thought for a moment, looked at Watson, then spoke. "Let's go for the entire Indian Ocean."

In Victoria the pair put their points to the President.

He listened and was very enthusiastic: we should strike while the iron was hot and work for success that June–July. He told us how, as a boy, he had stood on island shores and watched sperm whales and but now no longer saw them. We told him we knew the Soviet whalers were killing big numbers as they traversed the Indian Ocean from the Gulf of Aden to and from the Antarctic every Southern summer.

We talked about the composition of the delegation to the IWC. We also agreed that the delegation would need experienced advisors on law, science and political campaigning. As I was about to retire and was on leave, I felt sure my superiors in FAO and UNEP would allow me to serve as science advisor.[5]

Sidney's extensive earlier work with the International Ocean Institute (IOI) in Malta during the Law of the Sea negotiations at the UN led him to focus on securing support for the proposal from Indian Ocean coastal states, whether or not they were IWC members. So it was decided the Seychelles government would consult all other Indian Ocean coastal nations before finalising the sanctuary proposal. Sidney quickly engaged well-connected ocean advocates in his network, including Dr Sylvia Earle and Sir Peter Scott, to add scientific expertise and enhance the credibility of the consultation process.

Thanks to the courageous work of Sidney Holt and others, the Indian Ocean Sanctuary was established at the London meeting of the IWC in 1979.

The Moratorium on Commercial Whaling

The question of a moratorium or ban on, an end to, a pause in, or a suspension of, commercial whaling was undoubtedly the most important issue at the 1982 meeting of the IWC. Sidney had thought long and hard about the precise wording of the Seychelles Moratorium proposal. He had of course also drafted it.

Japan's IWC Commissioners had consistently claimed that what they always referred to as "a blanket moratorium" would violate the Convention. Sidney and other proponents of the moratorium did not believe this but did not wish to provide Japan and other countries an opportunity to make that particular argument.

On Sidney's advice, Seychelles emphasised that the proposal was not for an irreversible ban and would come into force three years from the date of passage. The reason for a delay, and its duration, were linked not only with meat supply contracts and needs for time for socio-economic adjustments but also because the IWC had previously agreed on a number of multi-year block catch limits in the Northern Hemisphere, through to 1984 in the North Pacific but to 1985 in the North Atlantic. It would have been tactically difficult, if not impossible, to cancel those.

The amended proposal was adopted by twenty-five in favour, seven against and five abstentions. Amidst wild applause, particularly from the observers' table at the back of the room, and a staged walk-out by members of the Japanese delegation, Chairman Iglesias adjourned the session.

Many commentators have subsequently described the 1982 decision as a "ban" on (commercial) whaling. It was not. At that time any such proposal would not have attracted even a simple majority of Members' votes. Nevertheless, it would not have been adopted were it not for votes from a minority of countries that were clearly opposed to the continuation of commercial whaling, such as Australia, India, France and the UK.

Attitudes of former "pro-whaling" countries have since changed dramatically. The worldwide growth of whale-watching worldwide has helped. The ecotourism activity now takes place in 120 countries and territories introducing millions of people each year to living whales while bolstering coastal economies worldwide. As my friend and sometime IFAW colleague Vassili Papastavrou puts it, animals, people and coastal communities all do better when whales are seen and not hurt.

Thanks to the courage of Sidney Holt and others, the proposal to establish a worldwide Moratorium on commercial whaling was passed at the Brighton meeting of the IWC in 1982.

Twenty-four months later, the Government of Japan initiated a "scientific" whaling program in the Southern Ocean around Antarctica and in domestic and international waters of the North Pacific. Iceland and Norway ultimately continued taking smaller numbers of whales in their domestic waters.

Putting the Noose on Antarctic Whaling

Pour one quart of Holt and Sons award-winning olive oil into large pot, add copious amount of local red wine; mix in equal parts world's top fisheries biologist, planet's leading environmental campaigner and audacious French entrepreneur. Sprinkle liberally with l'audace and herbes de Provence. Cover. Heat politics and public support till well done. Serve.

When Sidney Holt, then with IFAW, Greenpeace International founder David McTaggart and French entrepreneur Paul Gouin started cooking, there was no telling what might be served up. One Spring evening in 1992 it was a new initiative to curtail Japan's ongoing "scientific whaling" by Japan in defiance of the IWC moratorium: creation of a Southern Ocean Whale Sanctuary.

Sidney's friend Jean Paul Gouin, who had been helpful behind the scenes in securing the whaling moratorium, visited Sidney and his longtime collaborator and soulmate Leslie Busby at their farmhouse in Umbria during the festive season of 1990. As they considered prospects for turning the IWC moratorium or "pause" in commercial whaling into a permanent ban. Gouin floated the idea of a step towards that end: to declare Antarctic waters as a sanctuary for whales, and the trio prepared the rough draft of a campaign proposal.

Early in the new year, Gouin went to Paris where he met with Environment Ministry officials. The Environment Minister in President Mitterrand's cabinet, Brice Lalonde, was enthusiastic about the Sanctuary idea as were a few of his associates.

In the Spring of 1992, good friends again gathered at the farmhouse including Greenpeace Founder McTaggart and the brilliant Australian IWC Scientific Committee member Bill de la Mare. In the course of the dinner conversation, de la Mare noted that before they were brought close to extinction by pelagic whaling probably four-fifths of the world's baleen whales fed in the Southern Ocean.

"What did you say the new IWC minke catch limit might be?" McTaggart asked casually. Bill told him and he choked on his wine. "Got to do something about that!" All gathered were aware of the French Environment Minister's interest in the Southern Ocean Sanctuary idea Gouin had presented the previous year.

Impulsively, McTaggart picked up the phone, and dialled Paris saying, "I'm going to talk to Brice."[5] He and Lalonde, friends since the 1970s, spoke briefly about the Sanctuary idea.

The next day McTaggart was in Paris and had convinced Lalonde that the sanctuary was winnable; Lalonde, in turn, issued a press release announcing that France would be presenting the proposal to the next meeting of the IWC, which would be held in Glasgow in June 1992. As luck would have it, it was Lalonde's last act in office; he resigned from the government the following day.

Sidney, the Seychelles and others involved had secured the Indian Ocean Sanctuary with a surprise move. The 1982 moratorium, by contrast, had been cooking for ten years and success depended on the rate at which whaling countries were dropping out of the business and some new countries were joining the IWC, partly as a result of the earlier Indian Ocean initiative.

The Southern Ocean would be a much bigger bite, which would, at least, need the support of as many of the Antarctic Treaty Consultative Powers as possible, and as many of the other Southern Hemisphere coastal states as possible – a tall order.

Because Sidney had experience helping draft the texts of the two landmark proposals – both put forward thanks to his close work with the government of Seychelles – he was asked to make the first draft the Government of France would present to establish the Southern Ocean Whale Sanctuary (SOWS).

The campaign began later in 1992 at the IWC annual meeting in Glasgow. IFAW whale biologist Vassili Papastavrou had been summoned to assist Sidney during the meeting. "I booked into a room next to Sidney's at the Railway Hotel. I was immediately struck by his huge energy and output," Vassili would later recall.

> One morning he banged on my door at 7:30 am with two letters to national newspapers that he wanted me to fax (responding to articles that had only appeared that day), and a 15-page technical briefing, written overnight, that needed photocopying and distributing to IWC Commissioners. Throughout my career with IFAW, I often found myself struggling to keep up with simply reading Sidney's prodigious output.

A massive public campaign followed, largely underwritten by IFAW founder Brian Davies who committed close to a million dollars in institutional support to the effort, which involved production of a major documentary feature, *The Last Whale*, helpfully aired by Cable News Network (CNN), at the direction of McTaggart's friend and CNN Chairman Ted Turner. In addition, a worldwide petition effort was also promoted by another McTaggart friend (and countryman) Bryan Adams, an internationally known Canadian rock star during his world tour in the lead-up to the 1994 IWC meeting in Puerto Vallarta, Mexico.

Unbeknownst to both of us at the time, Sidney and I began our own episodic collaborations in the run-up to the 1994 Sanctuary vote. Together with other funders, IFAW was providing generous philanthropic grant support to the US branch of the Global Legislators Organization for a Balanced Environment (GLOBE) of which I was director.

Reviewing briefings on the Sanctuary proposal prepared by Sidney, Leslie Busby and the IFAW whale team I would one day lead, I immediately alerted GLOBE USA President Senator John Kerry and a particularly well-positioned founder of the global legislators' group – Vice President Al Gore. Both longtime environmental leaders enthusiastically engaged to ensure active US support for the Sanctuary proposal.

Gore himself spent some hours on the phone with South American and Caribbean Heads of States, patiently locking down their votes, or abstentions, in the case of several Caribbean states unwilling to oppose Japan, to secure the supermajority the IWC measure would require. After the Sanctuary victory, with assistance from your author, IFAW purchased full page advertisements in *USA Today* and the *Washington Post* featuring a '*Saved* the Whales' lapel button above the words: '*This button belongs to Al Gore*'. Working with political advisors to then Argentine President Carlos Menem I arranged to place a similar advertisement in the Buenos Aires paper of record.

Thanks to the courage of Sidney Holt and others, the proposal to establish a Southern Ocean Sanctuary for whales was passed at the Puerto Vallarta meeting in 1994.

Burning Bright at Twilight

Throughout the quarter-century that followed, Sidney worked tirelessly to tighten the noose on commercial whaling, leading a spectacularly talented, if largely unheralded team of scientists and campaigners he had recruited to work with IFAW. He collaborated extensively with other NGOs concerned with marine conservation including IFAW, Greenpeace, WWF, Whale and Dolphin Conservation and the Third Millennium Foundation, and as scientific advisor to Sea Shepherd Conservation Society (SCCS).

As Dr Rodger Payne, himself a world class scientist/advocate wrote upon his friend and colleague's passing,

"He was the natural leader of most aspects of the movement to end whaling. His willingness to work with organizations and individuals that most other scientists lacked the courage or imagination to engage with NGOs. In my opinion, this was, in a very basic way, the most important facet, the secret weapon, the leading role that Sidney Holt played throughout the 60 years his contributions guided and dominated the Save-the-Whales movement.

Sidney's willingness to work with a group like IFAW had two major effects on conservationists:

1 It demonstrated memorably that the tut-tutting towards NGOs that so many scientists practiced was one of the bigger mistakes for a scientist to make. Sidney Holt led the vanguard that demonstrated what a major waste of intellectual resources such tut-tutting could be.
2 Sidney's work with NGOs reassured many scientists who had gone on for PhDs, but hadn't dared work for NGOs to apply—a step that also benefitted NGOs in unrelated fields, as more PhDs summoned the courage to join their staffs."[8]

If ever Sidney's courageous example were relevant, it is now, as the gathering crisis of global climate change begins to accelerate profound shifts on the ocean planet, especially in the marine environment on which the survival of whales and human welfare both depend.

Last Words

By the time of its 2009 annual meeting on the sunny island of Madeira, Portugal, the future of the IWC did not look bright. The forum was struggling to stay afloat amid a maelstrom of political machinations, most involving the Governments of the US and Japan.

Offered one opportunity to address the plenary, international NGO representatives gathered and unanimously elected Sidney to speak on our behalf. A hush fell over the plenary as Sidney made his way, in halting steps, to the microphone. His hand was unsteady, his voice clear and sure:

"Mr Chairman, Delegations and Observers,

My name is Sidney Holt

…I am honoured to have been chosen to address you by the NGOs here concerned with animal welfare, conservation and the environment.

We intend to be pro-active, not merely re-active.

We wish to focus on the future of whales and the ecosystems they inhabit, not just the future of the IWC. Still, we want the IWC to survive.

By the way, this is a multiple anniversary year. It's important to me because I first became involved with the IWC exactly half a century ago – 1959. It was decided then that Antarctic baleen whale catches would be reduced to sustainable levels, by 1964 at the latest, in accordance with scientific advice to be provided by three independent scientists of which I was one. But that reduction didn't happen until the early 1970s.

Then, 2009 is the thirtieth anniversary of the creation of the Indian Ocean Whale Sanctuary. Next month the first symposium on the cetaceans of the region will be held in The Maldives, attended mostly by young scientists from the region.

Most importantly it's exactly eighty years since the eminent Argentine international lawyer, José Leon Suárez, proposed to the League of Nations that a sanctuary for whales be established in the Antarctic. Suarez reported that if nothing were done the fin, blue and humpback whales would be practically exterminated in the Southern Hemisphere. That took rather longer than he thought it would, but it had happened by 1959.

Then the sei whale resource was plundered in the 1960s. Demolition of the minke whales was begun in the 1970s. The biomass of the still numerous minke whales is less than one percent of the biomass of the Southern Hemisphere baleen whales at the time Suarez reported to the League of Nations.

Think about that. We're talking endlessly about how to sweep up the crumbs left on the table after the feast. If anything's dysfunctional, that's it.

All the NGOs for which I speak unreservedly support at this time the continuation of the moratorium, with no arbitrary catch limits being set. But they think it's time to move on: to end all commercial whaling under unilaterally issued Special Permits, all whaling in sanctuaries, all whaling under objections. And all international trade in commodities from Appendix I CITES listed species.

An end is justified by the improvement of scientific knowledge about whales, using non-lethal methods, and by the increase in scale and extent of non-lethal uses of whales. Furthermore, increases in threats to the survival and welfare of whales – resulting from the intensifying and growing diversity of human activities in and around the ocean – mean that relieving the ecological stress caused by whaling is now even more urgent.

The wondrous, vulnerable whales will never contribute substantially to the food security of humans. Nor do they threaten it. Despite insistent propaganda they're not responsible for the troubles of the fishing industry.

Commercial whaling is now unnecessary, is inhumane, and is even unprofitable, continuing – subsidized – for minimal financial gain.

Nevertheless, we in civil society insist on being conciliatory and constructive. The three-year phase-in of zero catch limits after 1982 allowed six whaling countries to make the social and economic adjustments needed to fold their operations. That should be long enough now for a phasedown, and out, of residual commercial whaling. The catches in that period should be fewer than in recent seasons; no new whaling vessels should be brought into service, and no new whaling operations begun. The phase-down and -out should be fair to the whaling countries that did abide by the IWC's 1982 decision. Intransigence should not be rewarded.

...Some other useful things could be started during the phase-down and -out.

They include resuming negotiations for revising the ICRW and also launching more research – as promised to the United Nations in 1972 – on the recovery of the whale populations and ecosystems that were severely impacted by poorly regulated commercial whaling, as well as to gain more knowledge about the new threats to cetaceans.

"Those are our suggestions, from all six continents and many small island states, includ-ing from all whaling countries."[9]

Sidney passed away a decade later in December 2019, a year after he celebrated the end of Antarctic and High Seas whaling, achieved in no small measure due to his lifetime of extraordinary work. Upon his passing, ICES, the very organ-isation that had first initiated international efforts to conserve whales 70 years before, posted a memoriam:

If Sidney were with us today, we believe he would dare us to ask:

What assumptions are we making that remain unchallenged?

What systems, and what research and policy, do we take for

granted, as either unchanging or unchangeable?

Which missing factors must be considered, and what new perspectives might add value?

Such questions exemplify what scientists today may learn from his legacy. By consid-ering the broader perspective and by collaborating outside our core disciplines and scientific circles, we can gain greater breadth of knowledge and become better equipped to develop solutions. Further, if we are to address today's most pressing problems, we need to remain focused on practical solutions that are not only grounded in sound science but can also find real-world implementation and enforcement.

When we feel the proposed solutions are unfeasible or unsubstantiated, we should be empowered to speak out, while making sure that criticism is constructive and followed up with alternative solutions.

Any conclusions and solutions must be based on strong evidence and a high level of rigor. No matter how monumental a piece of work, Sidney challenged us to continue to appraise it so that it could be improved and to reflect on it within broader contexts as well as in light of new information.

Finally, we must find ways to document our past and integrate historical data and perspectives

to inform modern and future management, and so we may learn from our mistakes and the decisions we have already made.[10]

Like the torch he so long carried, Sidney's legacy of courage has now been passed to us.

"Come my friends, 'Tis not too late to seek a newer world." – Tennyson.[11]

Notes

1 Related by Leslie Busby, 2022. (Retelling the story years later Antonella Papini was still indignant, and rightly so.)
2 Email message from Sidney Holt copied to the author, July 2016.
3 Sidney Holt, In His Own Words YouTube video interview https://www.youtube.com/watch?v=np2Eep99ohg
4 Clapham, P. and Ivashchenko, Y. (2009) "A Whale of a Deception", Marine Fisher-ies Review, Volume 71 Issue 1. Available online at: https://spo.nmfs.noaa.gov/sites/default/files/pdf-content/mfr711contents.pdf
5 Holt, S., 2022, "Save the Whale! Memoirs of a Whale Hugger."

6 Holt, S., 2008, "Three Lumps of Coal: Doing Fisheries Research in Lowestoft in the 1940s." Talk to CEFAS Seminar.

7 Papastavrou, V., 2022 Postscript, "Save the Whale!" Memoirs of a Whale Hugger.

8 Payne, R., 2020, "Losing Sidney" Ocean Alliance Blog https://whale.org/losing-sid-part-1/

9 Holt, S., 2009, "Statement on behalf of IWC NGO Coalition" https://www.asoc.org/storage/documents/Meetings/IWC/61st/Final_Sidney_Holt_Statement_IWC_61.pdf

10 *ICES Journal of Marine Science*, (2021), doi:10.1093/icesjms/fsab019 https://www.researchgate.net/publication/349757162_Sidney_Holt_a_giant_in_the_history_of_fisheries_science_who_focused_on_the_future_his_legacy_and_challenges_for_present-day_marine_scientists

11 Tennyson, A.L., 1850, "Ulysses" from "In Memoriam" https://www.poetryfoundation.org/poems/45392/ulysses

3

MOSTAFA TOLBA

The Egyptian King: The Montreal Protocol

Chris Spence

The year 1982 is memorable for many reasons. In the West, the US and Britain struggle with high unemployment and social discontent, while in the East China's population exceeds 1 billion for the first time. Long-serving Russian leader Leonid Brezhnev dies and is replaced by KGB boss Yuri Andropov. The Falklands (Malvinas) War flares in the South Atlantic, pitting two unlikely foes— Argentina and the United Kingdom—against one another in a short but bloody conflict. Meanwhile, Israeli troops invade Lebanon, huge nuclear disarmament rallies take place in New York and Greenham Common in England, and anti-apartheid movements in the US and Europe grow, increasing the pressure for reform in South Africa. In sports, Italy wins the football (soccer) World Cup with a 3–1 victory over West Germany. And American pop star Michael Jackson releases Thriller, which becomes the biggest selling album of all time, eventually shifting more than 70 million units (and counting) worldwide.

Yet in retrospect, none of these events were arguably as important as one that failed to make any headlines at all. The meeting of a new UN working group on an obscure gas went virtually unnoticed. And why should it attract attention? The UN often sets up committees to work on one topic or another. Even in diplomatic circles it hardly registered as a big deal. What's more, there was a powerful coalition of countries and business interests which felt action on the topic was premature, if not downright unnecessary.

Yet in spite of this a handful of experts led by one energetic Egyptian were determined to fight for progress, convinced the task at hand was one of the most important the world had yet faced.

This small group turned out to be right. The issue in question was a threat to the Earth's ozone layer. And the man leading the fight was Mostafa Tolba.

DOI: 10.4324/9781003202745-4

Rebellious Beginnings

Mostafa Tolba was born in 1922 in Zefta, an Egyptian town nestled on the western banks of the Nile. For hundreds of years, Zefta had been an important trading centre, an entrepot where boats and camels, caravans and people travelled in every direction. By the 1700s, it was a major hub for trade in coffee and a multicultural melting pot welcoming people from far and wide.

But in 1922 that was all in the past. By the early 20th century, Zefta's story had taken a more violent turn. Just three years before Tolba's birth, the townsfolk of Zefta had risen in rebellion as part of the Egyptian Revolution of 1919. That uprising had been put down by British troops, although within three short years—the same year Tolba was born—the world's largest empire had recognized Egypt's independence.

Did the history of Tolba's birthplace influence his later life and personality? Certainly, Tolba was subsequently to display some of his hometown's characteristics: an ability to work with people of all countries and cultures, a fierce determination, and a strongly independent mindset. Tolba also didn't like being dictated to, especially by colleagues and partners from Britain and other former European colonizers. He himself could often be autocratic, much like the monarchs of old. In fact, it could be argued Tolba bore many of the hallmarks of royalty. He could be magnanimous and charming, mercurial and overbearing, even dictatorial at times. But like the greatest kings, he was also exceptionally determined and single-minded.

Interestingly, these days it is another Egyptian, a fleet-footed footballer named Mohamed Salah, whom fans call 'the king.' Supporters of Liverpool Football Club in England even sing a song from the stands touting the talents of their hero. The chorus goes like this:

> Mo Salah, Mo Salah, Mo Salah,
> Running down the wing
> Salah, La, La, La, La, La, La,
> The Egyptian King.

Perhaps Mohamed Salah deserves his 'kingly' reputation—especially for his off-field largesse in giving to good causes. However, I would argue Mostafa Tolba has an even greater claim to being labelled a modern 'Egyptian King'—as this chapter will show.

Of course, in the 1920s this was all far off in the future. As a boy, Tolba showed academic promise, eventually attending Cairo University and later attaining a PhD from Imperial College London. His career started well and in the 1950s he worked at Cairo University and at the University of Baghdad. Later, he was employed by the Egyptian government.

A Global Role

It was in government where Tolba first cut his teeth on the international stage. He served briefly as President of the Egyptian Olympic Committee before being appointed to lead his country's delegation to the United Nations Conference on the Human Environment, held in Stockholm in 1972. Among its many achievements, the Stockholm Conference created the UN Environment Programme and in 1973 Tolba was appointed Deputy Executive Director. Two years later, when Maurice Strong stepped down as its first leader, Tolba was promoted to the top job.

Immediately, Tolba set about putting this small, new UN programme at the heart of global action on the environment, espousing a belief in 'development without destruction.' This philosophy showed Tolba's awareness that environmental protection would require a new type of global diplomacy. First, it would need to be based squarely on the best scientific knowledge available. Second, it would have to be sensitive to the very different circumstances of the world's wealthiest nations and of poorer, often newly independent countries also aspiring to develop and grow. Tolba knew these countries wished to build up their economies and help their people escape poverty. At the same time, he was also eager to avoid more of the environmental destruction already caused by industrialization and population growth in the West over the previous century.

By the early 1980s, Tolba had already helped galvanize the international community to action on several topics, including launching an international education programme to raise further awareness of environmental issues, and helping strengthen and grow UNEP's Regional Seas Programme.

Tolba had also been actively promoting further research and monitoring of the ozone layer. As early as 1975, UNEP was supporting research and expert meetings on the subject. So why was this issue—which had not been on anyone's radar a few years earlier—now becoming of interest to scientists and policymakers?

An Atmosphere at Risk ... or a Storm in a Teacup?

Ozone is a rare gas. Pale blue in colour and with an unusual smell a little like chlorine or burned metal wires, it can be found throughout the earth's atmosphere, albeit in tiny amounts. The only place where it is exists in greater concentrations is about 15–35 kilometres above the Earth's surface in the lower stratosphere, where it forms what is known as the ozone layer.

Why does this matter? It turns out the ozone layer, which was not even discovered by scientists until 1913, plays a critical role in protecting life on Earth. It absorbs roughly 98% of the ultraviolet (UV) radiation emitted by the Sun. Since UV damages DNA and is harmful in large doses to most known life forms, it is probable most life on Earth could not exist without this protective layer in our atmosphere. In humans, too much UV radiation causes a range of ailments, from weakening our immune systems to eye cataracts and skin cancer.

The discovery of the ozone layer—important though it was—may have been of only modest practical interest were it not for two more scientific breakthroughs just 60 years later. By the mid-1970s, researchers had started to worry that human-made substances might be causing the ozone layer to weaken. Specifically, they started to suspect a family of chemicals known as chlorofluorocarbons (CFCs) might be breaking down and destroying ozone molecules.

The researchers were surprised. Up until that point, CFCs had been considered a scientific and commercial boon. Since their invention in the 1890s, they had gradually come to be seen as a safe chemical with all sorts of uses. CFCs are non-flammable, stable gases that are not toxic to humans. By the 1950s, they had proven their value as a coolant in refrigerators and air-conditioners, as well as a propellant in sprays for hair products and a host of other domestic goods. They were even seen as good solvents for telecommunications gear and a growing range of other industries. In short, they were firmly established as yet another wonder of modern science.

Reflecting this viewpoint, production of CFCs had risen dramatically, from under 100,000 tons in 1950 to almost 1 million tons globally by the mid-1970s.

But this miracle product had a potentially fatal flaw. In 1974, researchers in the US published two separate theories. At the University of Michigan, scientists Richard Stolarski and Ralph Cicerone hypothesized that small amounts of chlorine, if released into the stratosphere, could attack and destroy ozone molecules in large numbers. That same year, Sherwood Rowland and Mario Molina at the University of California, Irvine, suggested CFCs, which contain chlorine, carbon, and fluorine, could last in the atmosphere for decades before being broken down into their constituent parts by solar radiation. When they do finally break down, however, they would release large amounts of chlorine into the stratosphere.

Taken together, these two findings had disturbing implications. If CFCs were releasing chlorine into the atmosphere, and if chlorine destroys ozone, did that mean the ozone layer—and life on Earth as we know it—would be at risk?

The Battle Begins

Many US scientists responded rapidly to these two early reports, with further research and modelling by experts from both academia and the government—primarily in the National Oceanic and Atmospheric Administration (NOAA) and NASA. Could they confirm the theories about a link between chlorine and ozone?

At that early stage, there was little actual evidence to support the idea. Ozone exists only in small amounts even in the stratosphere, and concentrations can wax and wane by region and season. Although measuring ozone levels is possible and had in fact begun several decades before, the early numbers did not point to a reduction in ozone.

What's more, powerful big business interests were sceptical, arguing there was not yet enough evidence to act. 'The chlorine-ozone hypothesis is at this time purely speculative,' said an executive with Du Pont, which by that time was a major producer. His boss, Du Pont's Board Chair, had reportedly been even more outspoken, dismissing the theory as 'science fiction' and 'utter nonsense.'

Du Pont was not alone. The UK government also took a cautious position, possibly influenced by industrial manufacturers of their own such as Imperial

Chemical Industries (ICI). At that time, the British had a strong voice within Europe. They influenced others, like the French, who had manufacturers of their own and also seemed unconvinced of the need for rapid action. With such powerful opposing forces ranged against them, what possible chance was there that urgent steps would be taken?

Yet Tolba was determined. As a scientist, he was convinced by the theory's logic. He believed it was much safer for both people and the planet to err on the side of caution by taking action sooner than later. For him, the risks were just too great to do otherwise.

Tolba was not someone who would take no for an answer. Like a benevolent dictator or monarch, he could be autocratic—even imperious—at times. His management style was known to be top down and directive. Certainly, his leadership was of the old school variety. Even with those who did not work for him—such as senior government officials over whom he had no authority or control—he could be assertive and forceful if he felt strongly enough about something. And the ozone threat was, he believed, important enough to fight for. Systematically, he began pushing and debating with key governments on why they should take the issue more seriously.

In 1977, UNEP held an international meeting in Washington, D.C., to promote greater urgency. Tolba also set up a Coordinating Committee on the Ozone Layer, which began bringing together experts from around the world to examine the risks. The US government was interested and sympathetic, hosting a second meeting a month later.

Yet progress was slow. With heavy hitters like the UK and French governments, plus major manufacturers, still unconvinced, the international community was unable to agree on anything beyond bland statements of good intent. There seemed no chance of achieving the sort of serious global action and regulation of CFCs Tolba believed was essential. What's more, in 1981 a new US government was elected that had a strongly anti-regulatory ideology. Would it block efforts to impose restrictions on the international community?

So while the scientific case for action gradually become more compelling, the forces arrayed against Tolba and his allies appeared to be getting more powerful, not weaker.

Still, Tolba would not give up. Like the best king or general, he could be brilliant and inspired, adapting to changing circumstances and switching his battle plan in a moment. If his authoritarian instincts and bossiness were not working, he could turn on the charm instead—cajoling, coaxing, and flattering those he wished to persuade. So Tolba continued to engage with and praise the US government while also reaching out to others. Meanwhile, he also kept convening meetings of experts and gatherings of government officials, using every platform he could to ring alarm bells on this issue.

'Under the dynamic leadership of Tolba, UNEP was active from the beginning in trying to sensitize governments and world public opinion about the danger to the ozone layer,' wrote US diplomat Richard Benedick in his seminal work, *Ozone Diplomacy: New Directions in Safeguarding the Planet*. In fact, as the prospects for international action seemed to dim in the early 1980s, it was Tolba and his UNEP colleagues who kept the flame of hope flickering. 'It is no exaggeration to state that it was UNEP that kept the ozone issue alive at this stage,' Benedick confirms.[1]

A King in the Wilderness

Undaunted by this glacial progress, Tolba launched a new initiative. UNEP's '*Ad Hoc* Working Group of Legal and Technical Experts for the Preparation of a Global Framework Convention for the Protection of the Ozone Layer' was quite a mouthful. Still, it set out Tolba's ultimate goal: an international, legally binding treaty that could make a real difference.

Yet the group started badly. At that time, few governments were giving the ozone threat the seriousness it deserved. According to Benedick, ozone had become a 'low priority.' As a result, attendance at its first meeting in 1982 was poor and progress painfully slow, with delegates stumbling from one disagreement to another. The US now seemed less engaged, while in Europe the French and British continue to resist any dramatic change that might harm their manufacturers. Since in the European Community there was an emphasis on reaching consensus, this often meant a 'lowest common denominator' approach where the more cautious voices won out. This was certainly the case with ozone, where the European position—guided by the UK and France—remained unwilling to commit to cutting CFC production without further evidence. The Soviet Union and Japan, too, appeared unsure about the need for rapid action. Were Tolba's dreams of a breakthrough doomed?

Finding Allies

Still Tolba persevered. In 1983, his hopes were boosted by two events. First, a coalition of like-minded countries was formed that was committed to strong action. Known as the Toronto Group after the city where they first met, Canada, Finland, Norway, Sweden, and Switzerland began to collaborate together in support of Tolba's dream. What's more, the US had a change of heart under a new Environmental Protection Agency (EPA) administrator. In fact, the US was among a handful of countries that had already made significant progress in reducing domestic CFC emissions. Although some within the Reagan administration did not favour global regulation on ideological grounds, their voices were being gradually drowned out by other, more forward-thinking staffers. At last, Tolba's list of allies was growing.

In March 1985, Tolba and his partners managed to achieve something of a breakthrough. At a meeting held in Vienna, governments signed off on a new treaty—the Vienna Convention for the Protection of the Ozone Layer. Finally, there was a UN agreement that formally created obligations on the part of governments to take the issue seriously.

Unfortunately, though, the treaty itself was weak. With some countries still dragging their feet, the idea of significant cuts to CFC production did not end up in the final agreement. The text of the Vienna Convention did not commit governments to much more than taking 'appropriate measures'—whatever that might mean! It did not even define which chemicals were ozone-depleting. CFCs were just listed in an annex to the treaty, along with many other chemicals, both dangerous and benign.

Tolba and allies like Winfried Lang, an Austrian who chaired many of the negotiations around that time, knew more was needed. Still, there were some causes for hope. The Vienna Convention may not have done enough to stop the problem, but it was a start. For instance, it included language promoting more international cooperation and the exchange of data. More important still, it called for more negotiations and aired the idea of a legally binding future protocol—or additional agreement—that would actually regulate and reduce ozone-depleting substances. The stage was now set for the next—and most important—round of discussions.

Could Science Be King?

As UN negotiations continued, the scientific case for action grew. In May 1985, scientists with the British Antarctic Survey found major depletion of the ozone layer in the Antarctic, shocking the international community and the public and leading to talk of an 'ozone hole.' A year later, a new report by UNEP and the World Meteorological Organization, collaborating closely with NASA, found that CFCs could potentially reduce the ozone layer around the world by one-tenth over the coming decades. This meant dangerous increases in UV radiation would be striking earth with potentially devastating consequences everywhere, not just over the polar regions. The study also found it was not just CFCs that could destroy ozone. For instance, another group of substances, halons, had enormous ozone-depleting potential. Since these were being used widely in fire extinguishers and other products, it presented yet another challenge while also showing CFCs weren't the only culprits. It was now glaringly obvious to many people—and not just scientists like Mostafa Tolba—that the world needed to come together and agree to cut its production of CFCs and other ozone-depleting substances.

The Road to Montreal

And now began the hardest chapter of all. The international community had agreed to try to finalize a new treaty—a protocol—by 1987. Still, countries such as the UK, France, and Japan were resisting any outcome that would be too restrictive on their manufacturers. They preferred a softer agreement that would give industry a lot more time to adjust. But this was not what Tolba—or the science—demanded.

The Egyptian's hopes were boosted when the US finally seemed to put any ideological doubts behind it. By 1986 the Reagan administration, driven in part by Secretary of State George Schulz, recognized the compelling science and research of the past few years. Even the President himself came out strongly in favour of action. Did the fact that Reagan personally had two brushes with skin cancer around this time influence him at all? Was it instead the alarming news of an Antarctic 'ozone hole'?

Whatever the case may be, Tolba was now working even more closely with his US friends, who started to apply diplomatic pressure on other countries, particularly in Europe. Their goal was to peel away other European nations from the Anglo-French position that had dominated the continent to that point.

My friend Andrey Vavilov, a former senior Russian diplomat who later worked for UNEP, recalls a curious incident from this time. At a meeting on ozone protection in London he was introduced to British Prime Minister Margaret Thatcher. Andrey remembers how, without entering into any of the usual social niceties or asking for his opinion, she fixed him with her steely gaze and pronounced, with apparent sincerity and conviction, 'We must *do something* about these old refrigerators!'

A chemist by training, Thatcher must have known the evidence by then was quite compelling. However, there was still no sign the British were going to budge. 'The United Kingdom was the dominant voice on ozone [in Europe] ... [and] consistently opposed strong international controls,' recalls Richard Benedick.

In spite of this, Tolba and his American friends began to create cracks in the European position. The US had particular success with the Belgians, while others like the West Germans also seemed increasingly motivated to act. Meanwhile, New Zealand became more vocally supportive under an environmentally progressive Labour Government that had won office in 1984.

Tolba now tried to widen his list of allies still further by reaching out to developing countries. As an Egyptian he was himself from the Global South. From a country at the crossroads of many civilizations, Tolba actively engaged with leaders from Africa, Asia, and Latin America, as well as Islamic neighbours in North Africa and the Middle East. At first many were uncommitted or uninterested. After all, why did ozone production matter for non-producing countries like them? But with the US' diplomatic heft and Tolba's indomitable energy, they were eventually able to bring many nations, from Argentina and Brazil to Kenya, to join the 'yes' camp.

But still some of the Europeans and a handful of others held out.

Tolba's Time

Up to this point, Tolba's praise and flattery of the US, combined with his 'cajoling' and persuasiveness with other countries behind the scenes, had helped grow the alliance. But in the face of Europe's continued intransigence, Tolba was forced to try a new approach. At a meeting in Geneva in March 1987, he stepped out from what had often been a behind-the-scenes role and gave an impassioned speech to the assembled group, telling them the evidence was now absolutely clear and he would no longer allow anyone to pretend there was any serious disagreement left among the scientific community. After laying down this challenge to his opponents, he also launched another new idea, working with allies on an unofficial text for a strong new treaty, drawing in large part from American ideas for how a future protocol might look.

Tolba also took a strong line against the UK. Channelling his 'inner monarch,' he vigorously countered British efforts to soften the unofficial text, lecturing and browbeating senior diplomats who stood in his way.

'Bullying, cajoling, wheedling and threatening,' is how Maria Ivanova described Tolba in her informative book on the history of UNEP.

Would his approach work? As delegates gathered in Montreal, Canada, in September 1987, tensions were high. More than 60 governments were now involved in the negotiations, including a growing number of developing countries. Non-governmental organizations had also become increasingly engaged and the global media and public were, too.

What followed were eight days of intense talks. Tolba and conference chair Winfried Lang worked tirelessly, pushing the more reluctant countries to accept a strong outcome. Tolba was forced to use every weapon in his considerable diplomatic arsenal, from 'flying kites'—that is testing new language or ideas to see what might stick—to aggressive persuasion behind the scenes in an effort to get his way. His ability to think on his feet in search of solutions that would find widespread support was invaluable—as were his eloquence and passion.

'He brilliantly managed the ozone issue [by]... bullying the producers nations to cut back on CFCs [and]...ensuring the funds were there' to help developing countries, recalls Stanley Johnson, the current British Prime Minister's father, who was active in the environmental world for many years.

'Known (behind his back) as the Ayatolba ... [he would use] threats, dramatic challenges and [an] overwhelming presence,' recalls British negotiator Fiona McConnell from her encounters with Tolba. His 'forceful and persuasive arguments and wheeler-dealing' were notorious, she adds.

Finally, after days of almost non-stop discussions, Tolba's persistence was rewarded and delegates signed off on a deal. It had taken more than 12 years since the issue had first garnered scientists' attention, and five since Tolba had launched his UNEP working group. Nevertheless, the UNEP chief's kingly virtues—from his magnanimity to his imperiousness and his incredible tenacity—had finally paid off.

Tolba's persistence was soon vindicated. Within months, even more sceptical or cautious countries had recognized the importance of the issue and changed their position as the scientific case continued to grow. In fact, from the late 1980s onwards, the British government, in particular, took on a leadership role.

'The evidence is there,' Prime Minister Margaret Thatcher told colleagues in a major speech delivered at the UN in 1989, where she urged greater action on ozone depletion. Under Thatcher and her successors, the UK began pushing for stronger measures, including funding to support developing countries, something it has continued to do over the past three decades. Furthermore, the fears raised by big business proved to be solvable, as Du Pont and some others quickly developed alternative, non-ozone depleting technologies to replace CFCs, moving rapidly to dominate the global market with their new products.

But Tolba had also achieved something beyond ozone protection. He had shown how science could bring about important policy changes even at the global level, where agreement between dozens of countries was required. The Montreal Protocol was a remarkable achievement—a forward-looking, life-saving treaty achieved against the odds.

The Montreal Protocol

So what, exactly, did the Montreal Protocol do? First, it obliged countries to phase down the consumption and production of CFCs and halons, as well as to report back their data—that is, to measure how well they were doing in meeting their promises. Second, it established that both wealthy Western countries and the developing world would take on responsibilities, while acknowledging that less affluent nations would need financial support and more time to achieve their goals. This eventually resulted in a 'Multilateral Fund' to give technical and financial help to developing countries so they could also cut down on ozone-depleting substances. In addition, the Montreal Protocol established an ongoing process where countries could continue to review emerging science and strengthen their pledges as needed.

The results were rapid, with production of CFCs falling by more than half between 1987 and 1992. Even with this ongoing success, however, Tolba and his allies were not satisfied. As the evidence continued to grow, Tolba continued to press for greater ambition. In 1990, an additional agreement, known as the London Amendment, hastened the phaseout of CFCs, halons, and carbon tetrachloride, another ozone-threatening substance. In the years to come, the treaty was amended and strengthened five more times.

Surrendering the Crown

And what of Tolba himself? After leading UNEP for 17 years, the Egyptian finally stepped down from his role in 1992. He left a legacy of achievement of which anyone could be proud. The success of his tenure as leader was unquestionable.

Tolba's departure did lead to some speculation, though. Was the great Egyptian really ready to retire? In Stanley Johnson's version of events, the qualities that had led to Tolba's success were also, ultimately, responsible for his downfall. Johnson claims the imperiousness, bullying, brilliance, and hectoring that had brought Tolba such success in the 1970s and 1980s had by the 1990s simply made him too many enemies.

British diplomats involved in negotiations with Tolba in the early 1990s 'wondered if he had modelled himself on the Chief Whip in the British Parliament,' recalls Fiona McConnell. The Chief Whip is known as a strict 'enforcer' who coaxes, dictates, and if necessary, browbeats parliamentary colleagues on how they should vote. As a result, they can be feared and even disliked. A number of colleagues—perhaps more in Europe than elsewhere—were either bruised by their encounters with Tolba or perhaps jealous of Tolba's success. Stanley Johnson recalls that an organizational review was concocted by UNEP member governments whose ultimate purpose was to dislodge Tolba from his throne. Eventually, they succeeded.

Did Tolba's kingly virtues also prove his undoing? In other words, did he jump, or was he pushed? At almost 70 and after 17 years leading UNEP, Tolba may well have felt it was time to step down. Still, it is interesting to note that UNEP's governing body has never appointed anyone quite like Tolba to the director's seat since, and that four of his five successors have been Europeans.[2]

Whatever version of events is true, Tolba's legacy is remarkable. What's more, Tolba continued his pioneering efforts even after leaving the UNEP hot seat. Back in Egypt, he established the International Center for Environment and Development (ICED) and was on the boards of environmental organizations in the US and Egypt. He continued to attend UN meetings into his 90s and remained an influential and respected figure. He also continued to write, ultimately publishing close to 100 papers on environmental issues. He was recognized by many academic institutions, governments, and other groups around the world with awards and honours from Russia to Canada, Hungary to Morocco, recognizing his status among the pantheon of environmental royalty. He died in 2016, aged 93.

A Safer Future

Today, thanks to Mostafa Tolba and other pioneers like Richard Benedick and Winfried Lang, the Montreal Protocol, London Amendment, and other more recent additions are having a profound impact. The ozone layer is expected to fully recover by the 2060s and the risk of ever-greater UV radiation striking earth has been averted.

The impact of this cannot be overstated. Without these early actions, we would now be experiencing millions more cases of melanoma and other skin cancers. According to research published in 2020 by the EPA, a staggering 443 million cases of skin cancer, and 2.3 million deaths, will be averted in the US alone by the end of the century thanks to the Montreal Protocol.

The Montreal Protocol has also kept a lid on the number of eye cataracts, as well as reducing the impact on other species in any number of ways, from averting damage to plant and animal DNA to impaired photosynthesis in trees and food crops. Some of this harm to our plant and animal life would almost certainly have harmed humanity in very serious ways, including affecting food supplies. Thankfully, we have avoided this frightening prospect.

What's more, the Montreal Protocol has also had a positive impact on climate change, since almost all CFCs are powerful greenhouse gases.

Still, our efforts to protect the ozone layer have not ended. Even today, more action is needed. For instance, we now need to limit and monitor the use of hydrofluorocarbons (HFCs). These are substances used as a substitute for CFCs, but which are also potent greenhouse gases. A recently agreed addition to the Montreal Protocol known as the Kigali Amendment aims to tackle this problem.

While the work must continue, however, we can breathe much easier knowing the global community came together early and effectively to tackle the ozone problem, and that it continues to take it seriously today.

Notes

1 Benedick, Richard E., 1998, *Ozone Diplomacy: New Directions in Safeguarding the Planet*, Harvard University Press, p. 42. Benedick's detailed account of the ozone negotiations of the 1980s and early 1990s remains the most important and authoritative yet written, perhaps in part because Benedick himself was heavily involved, playing a valuable role as a US negotiator. While some feel the book is too positive about the US and too severe on other countries, its insider account makes it a 'must read' for those interested in learning more about the Montreal Protocol.
2 Since Tolba, UNEP has been run by a Canadian (Elizabeth Dowdeswell), a German (Klaus Töpfer), a German-Brazilian (Achim Steiner), a Norwegian (Erik Solheim), and a Dane (Inger Andersen).

Background References

Benedick, Richard E. (1998) *Ozone Diplomacy: New Directions in Safeguarding the Planet*, Harvard University Press.

Ivanova, M. (2021) *The Untold Story of the World's Leading Environmental Institution: UNEP at Fifty*, MIT Press.

Johnson, S. (2015) *Stanley, I Resume: Further Recollections of an Exuberant Life*, Robson Press.

McConnell, F. (1996) *The Biodiversity Convention – A Negotiating History*, Springer Press.

Litfin, K. (1994) *Ozone Discourse: Science and Politics in Global Environmental Cooperation*, Columbia University Press.

4

FRANZ PERREZ

Putting the 'Basel' Back into the Basel Convention

Craig Boljkovac

Very few government employees are so committed and passionate that they are able to influence the agenda of their own country and to convince their politicians to take decisive action – let alone persuade the global community of governments, too. This chapter examines just one example of the commitment and influence of such an individual, Franz Perrez. In fact, it was difficult to choose only one example, considering the breadth and scope of his efforts over years of unceasing activity, which include the relationship between the World Trade Organization (WTO) and the environment, environmental governance, chemicals and waste policy and climate change.

When one thinks of international transport one usually thinks of the highly developed trade routes (or 'supply chains') that bring goods to market to the farthest reaches of our planet. In our interconnected world, however, there is a darker side to transport across national boundaries. There is a long history of waste flows from developed countries to developing countries – many of which do not have the capacity to handle what they receive. There have also been instances of 'donations' of, for example, computing equipment that are very close to the end of their useable life (or even beyond their useable life) to developing countries. These 'gifts' result in piles of electronic waste that the receiving country cannot safely process, because the computers or other goods are next-to-useless when they are received.

This sort of waste trade 'solves' some problems for developed countries, but in the end no one wins. Our planet is far too interconnected and dumping waste or near-waste also impacts the wealthier countries that do it in terms of their reputations as environmental protectors. In some instances, the pollution and resultant effects on the environment and human health caused by dumping in developing countries can cause significant, negative effects across borders. For

DOI: 10.4324/9781003202745-5

instance, some chemicals that are released can easily migrate on wind and water currents all over the world, much in the same way carbon dioxide becomes a global pollutant when it is released.

As a result, the international community has tried to sharpen some of the instruments at their disposal to ensure that such dumping is stopped in its tracks. Among the most targeted actions countries have taken to date is the Ban Amendment to the Basel Convention. One person, in particular, has done more in recent years to make sure this fundamentally important amendment has come into force as international law. This person is Franz Perrez, the Swiss Ambassador for the Environment.

From Lucerne to the World

Franz Xaver Perrez's origins are in the Canton (province) of Lucerne, in central Switzerland. He grew up in an 'ordinary' middle-class background in the Canton of Bern in a family with high environmental awareness. Since his college years, he has been passionate about music, literature, arts, politics and the environment. His additional passions for visiting and hiking the beautiful, almost-pristine Swiss Alps continues to this day, and his high-energy and enthusiastic approach to his life and work, which commenced in his early years, continues to be a driving force for his success as an environmental hero.

He has a modest approach to life: in his 50s, he has not yet learned to drive a car (an important conviction for a climate change negotiator) and in his teens, he refused to enter to an officer-track position despite being pressed by his military superior during the compulsory Swiss military service – he preferred his military service to be among ordinary enlisted personnel, with no special position because of his education or background. These life-long convictions have also served him well as an environmental hero.

Franz entered law school at the University of Bern, Switzerland, in 1985. Before completing his degree and passing the bar exam as attorney at law, he undertook graduate studies in Paris (at the Université de Paris II, Faculté de droit). As he travelled down his educational road, he increasingly specialized in areas of the law related to international negotiations. He achieved a Master of Law degree at NYU in New York, focusing on public international law, international trade and environment law. His PhD thesis 'Cooperative Sovereignty: From Independence to Interdependence in the Structure of International Environmental Law' focused on the principle of sovereignty as a concept of cooperation – something that, in retrospect, likely greatly helped him in navigating the complicated and nuanced world of diplomatic negotiations addressing potentially world-changing environmental issues.

Franz started his professional life in the Department of public international law in the Swiss Foreign ministry immersing himself into maritime issues. Representing Switzerland on the legal committee of the International Maritime Organization (IMO), and on the Central Commission for Navigation on the Rhine (the world's oldest international organization still operating; it dates from the Congress of Vienna in 1815). While these assignments might seem a bit technical or mundane, they helped Franz build the experience he would draw upon so effectively in later years when tackling burning issues such as the Basel Convention's proposed Ban Amendment. He gained further experience working for the Swiss economic ministry (SECO), representing Switzerland in the Committee on Trade and the Environment of the World Trade Organization (WTO). In this position, he developed and shaped the understanding that the relationship between WTO and environmental law should be guided by the principles of no hierarchy, mutual supportiveness and deference (that trade and environment are mutually supportive, and that one does not trump the other). In 2001 Franz joined the Swiss equivalent of an environment ministry: the Federal Office for the Environment (FOEN) International Affairs Division. With his work at FOEN (which has become more senior and important over time), Franz has developed a reputation as a top environmental negotiator and a key contributor, at least in the environment field, to the reputation Switzerland has as a country that punches well above its weight in international affairs. In addition to his work as environmental diplomat and negotiator, he also finds the time (since 2008) to teach international environmental law at the University of Bern's School of Law.

My personal engagement with Franz commenced in the early 2000s: I was a senior officer at the United Nations Institute for Training and Research (UNITAR), a small UN agency that had, at the time, a growing programme on Chemicals and Waste Management. Switzerland was competing with Germany to be the home of the Secretariats (the staff) of the Rotterdam and Stockholm Conventions (which, like Basel, deal with the global management of chemicals and hazardous waste). Franz approached me at an international meeting, and, with his characteristic smile and bright eyes peering at me through round, wire-rimmed glasses, indicated that Switzerland would be willing to support UNITAR projects to assist countries with capacity building related to the Stockholm Convention. From this point on, I was able to witness first-hand Franz's spectacular energy, drive and strategic insights into the world of multilateral environmental negotiations. We have been fast friends and colleagues ever since.

The Basel Loophole

Closely related to the Stockholm and Rotterdam Conventions is a third, older treaty: the 'Basel Convention on the Control of Transboundary Movements of Hazardous Wastes and their Disposal'. Negotiations for the treaty were finalized in 1989 and the convention entered into force (that is, became formal international environmental law) in 1992. However, this key convention for dealing with waste management on a global basis was widely considered a bit of a 'problem child' among multilateral environmental treaties (MEAs).

One of the main objectives of the convention is to prevent the transfer of hazardous wastes from wealthy, developed countries, to developing and/or least developed countries. The convention arose from actions in developed countries in the 1970s and 1980s, in particular, to restrict, ban and/or greatly increase the costs of disposing of hazardous wastes domestically. This drove facilities (usually in developed countries) that generated hazardous wastes to send them offshore – often to the very poorest countries (known in UN-speak as Least Developed Countries or LDCs) or to other developing countries. Upon arrival, and often for a fee that presented an enticing source of income for the recipient, the waste was dumped – usually untreated or with minimal treatment that failed to protect the environment and people from ill effects.

The Basel Convention entered into force in 1992. It established a so called 'Prior Informed Consent (PIC) procedure': hazardous wastes could only be exported if the importing country provided its prior informed consent. Thus, the exporting country has to inform the importing country about the planned export and about the hazards and risks of the wastes it wants to export. Export can only take place if the importing country consents (in writing) to the import. Several countries felt, however, that this procedure does not provide sufficient protection for developing countries and therefore proposed a general ban of exports of hazardous wastes from developed countries to developing countries.

Moreover, the Basel Convention applies only to countries who have ratified the convention. Although it contains a general provision prohibiting the transfer of hazardous wastes between parties and non-parties, one gaping hole has remained as the international community made efforts at implementation. The United States (US) has yet, to the present day, become a party to the treaty. This is particularly important as there is an exception in the convention: transfers between parties and non-parties can take place if the wastes are addressed in another 'convention' (usually a bilateral agreement between two countries) that does not 'take away' from the Basel Convention. The US has negotiated agreements such as this with an array of countries, including developing country parties to the Basel Convention. The failure of the US to ratify Basel is the 'thin edge of the wedge' – it has allowed numerous 'back doors' to circumvent the objectives, spirit and intentions of the convention.

Will the US become a party to the convention in the future? Probably not (while activists and others push for US ratification, the chances are slim). This failure to ratify is partly because of the complicated, and usually unsuccessful process, within the US for ratifying international conventions: the US Senate must confirm such ratifications, and, for a variety of political and usually internal factors within the US, such conventions are rarely ratified as they require two thirds of the Senate for this to happen. This is true not only for environmental agreements but virtually all international treaties. For the Basel Convention, this results in the world's largest per capita generator of hazardous wastes (at least by some measures) being technically outside a convention specifically designed to end their transboundary movement of these wastes – and actively circumventing it through side treaties.

As a consequence of these loopholes in the Basel Convention, recycling companies have exported almost vast quantities of hazardous wastes (according to the non-profit Basel Action Network (BAN), some 400 million tons per year). These exports include many things, including electronic wastes. When such wastes reach their destination in the developing world, instead of being recycled they are usually dumped and 'processed' using unsafe methods that expose workers, other local people and the surrounding environment to untold hazards as wastes are burnt, melted or otherwise 'treated' with hazardous chemicals. According to the Basel Action Network, for example some 40% of electronic waste delivered to US recyclers is exported to Asian and African countries.

In addition, other transboundary hazardous wastes issues continued to cause problems in developing nations and LDCs. Shipbreaking – taking apart huge container ships and other vessels when they reach the end of their useful lives – normally takes place in countries such as Bangladesh, China, India and Pakistan. These are countries with relatively weakly enforced environmental regulations. Such ships often contain hazardous substance such as asbestos and persistent organic pollutants such as polychlorinated biphenyls (PCBs) (highly regulated under the Stockholm and Basel Conventions). They pose huge hazards to workers who are dismantling such ships with minimal personal safety protections. Many of these activities are in probable violation of, among others, the Basel Convention, the Stockholm Convention and several treaties under the purview of the International Labour Organization aimed at the protection of the environment and workers' health.

Such gaping holes and persistently bad environmental and health practices clearly needed a solution. It was readily apparent that the Basel Convention did not go far enough to prevent the transboundary movement of hazardous wastes, despite many sincere efforts at the international level to negotiate an effective treaty.

When Is a Ban Not a Ban?

Thus, at the third Conference of the Parties (COP-3) in 1995, Parties to the Basel Convention adopted the Ban Amendment. The Ban Amendment, as passed, completely prohibits the export of hazardous waste from the Organisation for Economic Co-operation and Development (OECD) countries to non-OECD countries. The Ban Amendment provides for the prohibition of exports of all hazardous wastes covered by the Convention that are intended for final disposal, reuse, recycling and recovery from developed countries to all other countries. And it also bans exports from the USA to non-OECD parties of the Basel Convention having ratified the Ban Amendment.

The Ban Amendment was adopted despite the fact that several countries from all regions of the world (including both developed and developing countries) opposed such a ban. As a consequence, parties could not agree on the threshold for the entry into force of the Ban Amendment. What seems a technical legal question became a political issue blocking much of the important work of the Basel Convention.

To enter into force, two-thirds of countries that have accepted the amendment had to 'ratify' the amendment – that is, have it passed into law by their national lawmakers. While those supporting the Ban Amendment and its entry into force argued that this required the ratification of two-thirds of the Parties that were a Party of the Basel Convention when the amendment was adopted in 1995 (this 'fixed time approach' would require ratification by 66 countries who have been a Party of the Basel Convention as of 22 September, 1995), opponents argued that an entry into force required two-thirds of the current parties to the Basel Convention (this current time approach would require 126 ratifications). It was clear that the higher threshold could not be met. Moreover, there was acute, continuing opposition for ratification from an almost alarming array of important developed and developing countries, including Australia, Brazil, Canada, India, Japan, Mexico, New Zealand, the Republic of Korea (South Korea) and Russia. They were backed up by a key non-party, the US. Moreover, the division over the entry into force of the Ban Amendment over-shadowed much of the other important work of the Basel Convention. Now actively working in this area, Perrez and many others felt this situation was completely unacceptable.

A Tragedy in Abidjan

The determination of Perrez (among others) to change the situation had been strengthened in 2006 by a horrific incident in Cote D'Ivoire (Ivory Coast). A ship offloaded a highly hazardous liquid cargo which was dumped at many different sites in the Abidjan region. Estimates are that from 30,000 to 100,000 persons became sick, 17 died and dozens were hospitalized. The ship had tried to offload the cargo in The Netherlands (it apparently originated in Estonia and was treated on board the ship *en route*, resulting in even more hazardous chemicals being generated). However, once it was analyzed, the costs for proper treatment were considered to be far too high to be paid, and the liquid waste was actually pumped back on board the ship (an illegal practice under Dutch law). The ship then made its way to Cote D'Ivoire. In the aftermath of this incident, while compensation to some victims and other fines have been paid by the companies deemed responsible for these actions, many analysts feel that compensation was not at all sufficient. The one silver lining from this horrible event is that it proved to be a catalyst for further action to ratify the Ban Amendment (Figure 4.1).

As tragic as the Cote D'Ivoire incident was, it created an opening for Perrez and his allies to swing into action. The lack of ratifications of the Ban Amendment, coupled with this horrible, highly-publicized incident, provided needed momentum. Perrez felt strongly about the continuing, total blockage of the entry

FIGURE 4.1   Franz Perrez speaking at the joint Conferences of the Parties to the Basel, Rotterdam and Stockholm Conventions in May 2019. Photo courtesy IISD/ENB/KiaraWorth.

into force of the Ban Amendment. Not only because the Basel Convention was named after an important Swiss city that was the centre of its chemicals production industry (a relatively minor embarrassment), but, most importantly, because this dispute between Parties blocked much of the important work under the Basel Convention and prevented the convention from being effective and making the needed differences on the ground. He decided to take action to ensure the Ban Amendment would enter into force as soon as possible. His plan was to bring together a small but representative group of countries supporting and opposing the ban to take a less politicized and less emotional look at the Ban Amendment and the underlying problems. If such a process could show how the ban might work for everyone and could also offer options for those opposing the ban, Perrez felt, this might persuade countries to agree on the lower entry into force requirement for the amendment, providing the needed ratifications to bring it into force. Fortunately, Perrez found a willing partner for the pilot scheme – Indonesia.

Enter Indonesia...

Indonesia is a complicated and remarkable country. Comprised of thousands of islands – it is literally a huge archipelago – with a population of some 270 million persons – it holds a great diversity of cultures, languages and environmental and economic conditions, including among the highest levels of biodiversity in the world. Considering the challenges such a country faces, Indonesia has a solid track record of addressing environmental issues. However, as is the case with many developing countries, it has a long way to go to achieve the Basel Convention's goals of protecting the environment and human health. Weak environmental governance is an endemic problem – Indonesia ranks quite low in many indices addressing environmental performance. Thus, the entry into force of the

Ban Amendment would have helped Indonesia to protect itself more effectively from unwanted imports of hazardous wastes.

Just two years after the horrific incident in Abidjan, the ninth Conference of the Parties (COP) of the Basel Convention was held in Bali, Indonesia. With Indonesia hosting the meeting, Perrez felt this would help build additional support and momentum for the Country-Led Initiative (CLI). Moreover, a CLI with Indonesia on forest governance under the United Nations Forum on Forests (UNFF) had already been initiated (another process in which he was involved). He could therefore build on already existing good cooperation with Indonesia for such international environmental processes.

The COP in Indonesia was once again overshadowed by the lack of agreement on the Ban Amendment. After an effort to find a solution in a closed ministerial meeting failed, Perrez proposed that Switzerland and Indonesia launch a CLI on the Ban Amendment as a way forward. Backing up Perrez's instincts, Indonesia, as President, reaffirmed the importance of the Ban Amendment and supported the call for further exploration of how its entry into force could be achieved through a CLI.

An informal group was put together to explore what further measures could be taken. Perrez and his Indonesian counterpart, Emma Rachmawaty, were able to provide resources and a clear commitment to make it happen. Planning and further discussions on needs occurred at three meetings held between the 2008 Bali COP and the next (and tenth) Basel Convention COP, held in 2011 in Cartagena, Colombia. Key parties that both supported and opposed the Ban Amendment were convened by Switzerland and Indonesia, and proposals for a way forward were prepared for presentation at Cartagena.

During these meetings of the CLI, the group developed a better understanding of the Ban Amendment, namely that importing Parties (such as India) could still import hazardous wastes from OECD countries if they do not ratify the Ban Amendment (India and other developing countries had previously opposed the amendment because they wanted to continue being able to import such wastes for their recycling industries). It had also become clear, over time, that the majority of hazardous wastes imported to developing countries no longer came from developed countries, that the Ban Amendment thus had only limited effect and that therefore additional measures to promote the environmentally sound management of hazardous wastes were needed. And it highlighted the need for further legal clarity to distinguish wastes from products that could still be used.

'High Noon' in Cartagena

Cartagena is where the real drama began. The deliberations of the informal group on the CLI were, indeed, transformed into a series of recommendations for action for COP-10, but it did not start smoothly. On the night of the second day of the COP in what was perhaps a relatively early sign of the extreme weather incidents that now seem so commonplace as indicators of climate change,

Cartagena received a record level of precipitation. This truly set the stage for the negotiations as delegations struggled to reach the COP venue through knee-high water. Overnight floods led to a 'high-noon', as Franz termed it, at Cartagena.

The CLI proposals put forward at Cartagena comprised a draft, 'omnibus' decision, which included the following components: to understand the difficulties of the hazardous waste problem (including that the majority of transboundary movements were between developing countries of the Global South) and of the consequences of entry into force of the Ban Amendment (including that those not ratifying it would not be bound to it); and second, to come to agreement on additional elements that would be needed to make an entry into force possible (namely, to establish criteria for the sound management of hazardous wastes so that these could be used as criteria for the transboundary movement of hazardous waste by those countries who do not want to implement the ban). In a nutshell, the aims of the draft decision were to ensure that all (if any) hazardous wastes moving between countries had no adverse impact on the environment or human health.

The opening of the Cartagena COP included a key statement from the previous COP President and host, Indonesia, which called for, among other issues, the COP to resolve the legal interpretation of the convention on entry into force of amendments (recall that many opposing parties even disputed how many ratifications that would take). Ensuring sustainable financing for the convention and environmentally sound management of hazardous wastes were also highlighted.

This important statement did not occur in a vacuum. It was the culmination of over three years of close cooperation between Switzerland and Indonesia – and the fundamental building of trust that took place thanks to the efforts of key individuals (successful environmental negotiators, like superheroes, never act alone!). The statement was received by the COP with great effect. Perrez had also cultivated close ties and a true friendship with another committed environmentalist, the COP-10 President (and another environmental hero in this publication) Paula Caballero of Colombia. In her opening statement, she urged Parties to support the Swiss–Indonesian CLI.

Three years of intersessional negotiations, however, were not enough. Further serious negotiations had to occur in Cartagena for the CLI to be approved by governments. And this is where the full force of our hero Perrez's experience, commitment and drive went into action. First of all, after some generally supportive interventions, parties agreed to the formation of a 'contact group' – a small group that would commence the further negotiations, making recommendations for the plenary (all the parties present) to consider and hopefully adopt it. Another committed environmentalist and extraordinary diplomat, Osvaldo Alvarez of Chile, was chosen to chair the contact group. His heart and mind were also in the right place, and Osvaldo did an admirable job of steering the discussions in a direction that would benefit the environment and human health.

While the intersessional work had clearly bridged some of the gaps between parties who wanted the Ban Amendment to be ratified, and the significant number that did not, there were still major disagreements at Cartagena. Recognizing this, Perez made a critical introductory intervention in the plenary, stating 'Let the best not be the enemy of the good'. These words had great effect on the participants. Maybe a perfect outcome would not be achieved, but after almost 15 years of minimal progress towards the Ban Amendment's entry into force, even an imperfect CLI would be welcomed.

After tense, protracted negotiations, it was decided at Cartagena that the Ban Amendment would enter into force with three-quarters of the parties present and voting at the time of its adoption in 1995 – the so-called 'fixed time approach'. This was the far easier hurdle than the 'current-time' approach. Interestingly and to the benefit of all, the US played a positive role in supporting the 'fixed time approach' that made reaching an agreement easier, even though it was not, and is still not, a party to the treaty.

The *Earth Negotiations Bulletin* (ENB) —a well-known specialist publication that reports from many environmental negotiations, captured it best when it reported as follows:

> In the CLI contact group, long held resistance to the 'fixed time' approach appeared to have been swept away with flood waters. To many parties' surprise and praise, all seemed prepared to adopt the interpretation, which is set to facilitate the entry into force of the 1995 Ban Amendment.
>
> *(ENB, October 2011)*

Achim Steiner, the Executive Director of UNEP at the time, closed the Cartagena COP by proclaiming that it was the most successful Basel COP to date. With this needed momentum, the CLI continued, with the objective of assisting not only the 17 parties still needed to achieve entry into force of the Ban Amendment, but to provide assistance to all relevant parties who needed help. Participants were jubilant and elected Franz Perez as the next President of the Basel COP.

The Cartagena Conquest

What effects did the CLI have, once accepted at Cartagena? Franz worked with his Swiss government colleagues to ensure that it comprised assistance, including projects, that truly met the needs of parties who were struggling to have the proper capacity to ensure that the Ban Amendment would be properly enforced once it entered into force. The result? The rate of ratifications 'sped up', at least in a relative sense, although it still took a good amount of time to achieve the desired outcome.

Ten years later, in 2019, Perez's (and the world's) goal was finally realized when St. Kitts and Nevis, closely followed by Croatia, were the final two countries to complete the needed number of party ratifications for the Ban Amendment officially to enter into force. After almost 25 years, and thanks largely (but far from exclusively) to the incredible efforts of Franz Perrez, the Ban Amendment had become international environmental law. Finally, a much strengthened and effective Basel Convention came into being.

For Perrez, the entry into force of the Ban Amendment was a challenge asking for a concerted and sustained effort, strategic thinking and sheer force of will. How did this remarkable achievement come to pass, and how did one person have such a dramatic effect? How did such decades-long, entrenched resistance from many developed countries finally melt away? Many of his personal attributes seem to have been essential.

The Secrets of Success

While I was not personally in Cartagena to witness Perrez in action, I have seen him working hard at countless other COPs and international meetings. The attributes that 'make our superhero real', in my opinion, include: first, a good understanding of the issues – this includes understanding not only technical issues of substance, but also the political positions and the interests behind these positions. Second, taking a strategic approach that involves thinking beyond the next steps and beyond traditional positions and boundaries. And third, mapping out almost every minute of his time to maximize his effectiveness; meeting with strategic 'influencers' on a regular basis, and even keeping contact when there is a huge gulf between his country's position and theirs. Perrez seems to work morning, noon and night, with minimal sleep (this isn't recommended for everybody!). His incredible energy, combined with his humanity, kindness, humour and patience was a powerful combination. He maintained his compassion even when he seemed to be completely physically and mentally exhausted.

Perrez also counted on key allies, (in addition to the full support of his government) with whom trust has been built carefully over time. In addition to his Colombian, Indonesian and Chilean allies mentioned above, another key friend and ally was Jim Willis, the former Executive Secretary of the Basel Convention, with whom Franz was able to work closely despite their greatly different roles. Most importantly, Perrez was able to depend on the steadfast support of the African Group, of the progressive Latin American countries, the Nordic Countries and of the European Commission and European Union delegates, whose unequivocal, long-time support for the Ban Amendment was crucial. Franz could also rely on these allies and friends to achieve other important successes in the field of chemicals and waste policy, including the establishment of a joint secretariat for the Basel, Rotterdam and Stockholm Conventions, the adoption of the mandate by UNEP to launch the negotiations of the Minamata Convention, the listing of Endosulfan under the Stockholm Convention, or the adoption of the compliance mechanism by the Rotterdam Convention. Even when the gulf between positions seemed unbridgeable, Perrez never gave up in his search for common ground.

The Benefits of Boundless Enthusiasm

Perrez is among the most enthusiastic and positive persons that engage in these MEA negotiations. While international negotiations through the United Nations are often deeply adversarial and complicated, Perrez always tends to see the best in people, fully understands their positions and respects that they are defending the positions of their respective countries. He seems to usually be having fun during even the worst tensions or lowest points negotiations can reach, and he also seems to have been born into this field of work. All of these are attributes of a true environmental hero. If he has an easily identifiable flaw, it is his impatience (he is often too far ahead of others) – which can be displayed as irritability.

Finally, after many decades of effort, the Basel Convention was worthy of being named after a Swiss city, thanks, in part, to the unfailing commitment of our Swiss superhero. The 'Basel' was back in the Basel Convention.

In closing, I believe the Indian publication *Down to Earth* said it best when they summarized what took place:

> *With the Ban Amendment now a legal certainty, we hope the countries that have to date refused to ratify will do so and close the sad chapter of toxic colonialism done in the name of recycling.*
>
> *(Down to Earth, 2019)*

References

Down to Earth, (2019) Issue 10 September 2019. Available online at: https://www.downtoearth.org.in/news/waste/basel-ban-amendment-becomes-law-66651

Earth Negotiations Bulletin, (2011) Basel Convention (COP-10) Daily Report for 19 October 2011. Available online at: https://enb.iisd.org/events/10th-meeting-conference-parties-cop10-basel-convention/report-main-proceedings-19-october

5

RAÚL A. ESTRADA-OYUELA

Hero of Kyoto: The Kyoto Protocol

Joanna Depledge

The atmosphere was electric with expectation in plenary room II of the Kyoto International Conference Hall. It was 10:17 am on 11 December 1997, 16 hours after the Kyoto Climate Change Conference was supposed to have wrapped up. Removal staff were hovering impatiently outside, eager to prepare the site for the next event. This final night of negotiations to agree to the text of the Kyoto Protocol had started at 1:00 am and delegates were exhausted after 10 days (and often nights) of intensive work. And yet, while some heads may have drooped earlier in the evening, all delegates – governments and civil society observers alike – were now fully focussed on the front podium, where Chairman Ambassador Raúl A. Estrada-Oyuela sat in complete command. Hundreds of journalists and camera crews squeezed into every available space at the back of the room. And, for the first time in UN history, the meeting was being broadcast live through a ground-breaking new technology – the World Wide Web.

The eyes of the world were truly fixed upon Chairman Estrada as he banged his gavel and declared, using the time-honoured phrase, 'I hear no objections, it is so decided. The Kyoto Protocol to the United Nations Framework Convention on Climate Change is recommended for adoption'. And then, with a flourish, he added 'by unanimity'. This was no idle turn of phrase. The negotiations, over a gruelling two-and-a-half years, had been fiendishly complex, highly political, and often acrimonious. Just a couple of hours before, they had seemed on the verge of collapse. The President of the Conference, Japan's Environment Minister, Hiroshi Ohki, had to be dissuaded from retiring despondently to Tokyo, and the early morning newspapers in Japan were warning of a breakdown. Yet now, Estrada was able to proclaim not just consensus – as would be the norm – but *unanimous* agreement. He could be forgiven his sense of triumph.

DOI: 10.4324/9781003202745-6

The room erupted into loud and prolonged applause. Normally staid diplomats leapt up from behind their country nameplates to shake hands, even embrace, both allies and former rivals. Camera flashes lit up the chaotic scene like fireworks. Tears of joy, tiredness, and relief were on riotous display. In the overflow rooms, large screens broadcast the historic moment to cheering civil society observers. A few hours later, the Kyoto Protocol would be formally adopted by the Conference. Against the odds, the world's governments had united to place legally binding constraints on the greenhouse gas emissions that threatened to overheat the atmosphere and destabilise the climate. This was a moment for hope and rejoicing.

There are many forces that crystallised to enable this remarkable outcome. Undoubtedly, one of these was Estrada himself, who was acclaimed by commentators across the political spectrum. For Mark Mwandosya, the Tanzanian chair of the developing country Group of 77, 'His wit, sense of humour, persuasiveness, diplomacy and, when it was required no-nonsense approach, were qualities which, to a great extent, contributed to the success of the tough negotiations' (2000, p. 23). For David Sandalow, US Assistant Secretary of State, 'Estrada is a grandmaster of diplomacy and the godfather of Kyoto... It wouldn't have happened without his leadership, excellent judgment and good humour'.[1] Michael Zammit Cutajar, head of the UN Climate Change secretariat, summed it up in his official report to his UN bosses: 'I am convinced that, without his skilful and decisive leadership of the negotiations, a satisfactory outcome would not have been achieved'.[2] It wasn't long before Chairman Estrada would be dubbed the 'father', 'god father', and 'hero' of Kyoto.

The Kyoto Protocol negotiations: A prologue

To appreciate the significance of that historic moment in Kyoto, and the key role played by Chairman Estrada in making it happen, we have to understand the enormity of the task entrusted to him. It was in 1988 that climate change issue was first discussed in the UN General Assembly, triggered by a flurry of scientific warnings over the threat posed by rising emissions of greenhouse gases since the industrial era. Although the science was still emerging, projections of rising sea levels, soaring temperatures, and more extreme weather were sufficiently alarming to prompt the world's governments to launch talks on a new treaty aimed at combatting the problem. These early negotiations soon exposed the fault lines that would indelibly shape climate politics for decades to come: between advanced economies whose modern societies were built on fossil fuels, and developing countries fearing that the climate change agenda might threaten their own development aspirations; between vulnerable countries terrified of the impacts of weather extremes and rising sea levels, and oil exporters more concerned to avoid economic fallout from climate policies; between states and their leaders willing to contemplate a leap to a greener future, and those wedded to the status quo. And many other complexities besides.

The world's first climate change treaty, the United Nations Framework Convention on Climate Change (UNFCCC), was adopted on 9 May 1992, to scenes of jubilation not unlike those seen in Kyoto.[3] Just one month later, it was opened for signature by world leaders at the historic Earth Summit in Rio de Janeiro. As its name suggests, the UNFCCC set out an agreed framework for crafting an effective international response to climate change, establishing a set of principles, rules, institutions, and procedures that has endured over three decades. The treaty sets an ultimate objective of preventing 'dangerous' climate change, within a timeframe that would limit disruption to ecosystems, food supply, and the economy. To achieve this, the developed countries of the Global North (known as Annex I Parties,[4] because their names are listed in an Annex to the treaty) should 'take the lead' in curbing their greenhouse gas emissions, in line with the principle of equity and 'common but differentiated responsibilities and respective capabilities'. In other words, all countries share a common responsibility for the climate system, but the Annex I parties should act first, on the grounds of their greater historical contribution to climate change, their generally higher emissions country-wide and per person (at least in the early 1990s), and their stronger capacity to respond to the problem. In concrete terms, while all countries – from the poorest to the richest – assumed general obligations to respond to climate change and report on their emission levels, the Annex I Parties additionally agreed that they would aim to return their greenhouse gas emissions to 1990 levels by 2000, and supply more detailed data. The more advanced economies among them also committed to providing financial and technological support to the developing world.

It is easy to forget that, at the time, the division between North and South, poor and rich, was much greater than it is today. In 1992, for example, developed countries accounted for about three-quarters of global emissions (the US alone around one-quarter). China's GDP was roughly a tenth that of the US, and India's was one-twentieth. US emissions were almost double those of China, and seven times greater than those of India.[5] Chinese emissions per person were about half the world average, and those of each Indian citizen not even one fifth. The two-way categorisation of countries in the UNFCCC – much criticised in recent years – made perfect sense in 1992.[6]

The UNFCCC was a critically important first step in the global response to climate change. But by itself, it was not enough. The intent always was that supplementary treaties would add more ambitious and precise commitments. To this end, farsighted negotiators wrote in a 'trigger clause', which would oblige governments to discuss next steps as soon as the UNFCCC entered into force.

This discussion duly took place at the very first Conference of the Parties to the UNFCCC (COP-1), which opened in Berlin on 28 March 1995, under the Presidency of Angela Merkel, then Environment Minister for Germany. But countries were deeply divided as to the next steps to take. The most vulnerable states called for a new and stronger treaty to be adopted already in Berlin, whereas the oil producing countries objected to the launch of any new negotiations at all.

It took many nights of intense and acrimonious debate, and the personal inter-vention of Ms Merkel, before delegates finally agreed to decision 1/CP.1, which soon became known as the Berlin Mandate. This momentous decision launched a new round of talks to define stronger commitments for the Annex I Parties. In line with the UNFCCC's equity principle, this negotiating round would 'not introduce any new commitments' for developing countries, but all parties agreed to 'continue to advance the implementation' of their existing commitments un-der the Convention. A new forum was set up – the Ad Hoc Group on the Berlin Mandate (AGBM) – to conduct the negotiations, and Ambassador Raúl Estrada from Argentina was appointed to chair it. The negotiations that would eventu-ally lead to the Kyoto Protocol were underway.

Introducing our hero

The titanic struggle in Berlin foreshadowed the obstacles that would confront Chairman Estrada in the years to come. But if he was daunted by the Hercu-lean task ahead of him, he never showed it. A journalist and lawyer by training, Estrada was a career diplomat, enjoying a distinguished record with the Argen-tinian foreign service, including postings in Washington and Vienna. During the Kyoto Protocol negotiations, he also served as Ambassador to China, and would often joke that brokering trade deals on beef prepared him well for the rigours of climate change politics. Estrada certainly knew what he was letting himself in for, having been deeply involved from the very start of the global response to climate change, serving as Vice-Chair of the Intergovernmental Negotiating Committee (INC) that had negotiated the UNFCCC. He had then taken up the position of INC Chairman in the interim period before the UNFCCC's entry into force, brokering talks on the treaty's implementation details. At COP-1 in Berlin, he had honed his chairing skills by overseeing the debate on which country would host the UN Climate Change secretariat. This debate threatened to turn ugly, as Canada, Germany, Switzerland, and Uruguay all vied for the privilege. In a stroke of diplomatic creativity, Estrada persuaded delegations to hold 'an informal confidential survey' (a secret vote in all but name), from which Germany emerged the favourite. Not everyone thanked him for this, but at least the issue was definitively resolved. These experiences would serve him well, and built his reputation as a tough, ener-getic, and resourceful Chair. Estrada thus began his work in the AGBM already enjoying trust and respect. The UK's opening statement at the AGBM's first session echoed the general sentiment of the room: 'Many of us have sailed with you before, and it is therefore greatly reassuring to know that our helm is in your strong, experienced hands'.[7]

Speak to any Kyoto veteran about Estrada, and they will point to his strong personality, reflected in his rather rotund figure, well-kept moustache, and loud belly laugh. Blessed with supreme self-confidence and a thick skin, Estrada was indeed a force to be reckoned with. He could be charming, jovial, and impeccably diplomatic, but equally he did not suffer fools gladly and was unafraid of confrontation; he was 'the combative type',[8] as Zammit Cutajar put it.[9] A native Spanish speaker, Estrada was no great orator in English (which remains the *lingua franca* of global negotiations). His authority and leadership were founded not on fine words and stirring speeches, but on creative thinking, skilful political strategy, and a certain amount of vigorous cajoling. He was an interventionist and proactive chair, deeply engaged in every aspect of the negotiations. Most importantly, Estrada possessed great courage and did not shy away from taking difficult decisions, whatever the personal or professional cost. In an email to his UN secretariat support team as the negotiations entered their final stages, he pledged: 'I will do my best with the final outcome of AGBM, even at the cost of not being elected in any place in the world after the "crime"'.[10]

Chairing a global negotiation can be a lonely job. But Estrada inspired great loyalty from all those he worked with, and so, as he faced up to the demands of his role, he was able to rely on the unwavering support of his UN secretariat team (especially Michael Zammit Cutajar and Richard Kinley), as well as his right-hand man on the Argentinian delegation, Diego Malpede. His beloved wife, Leticia Vigil Zavala, also regularly accompanied him on the negotiation circuit and was, by all accounts, a profound source of encouragement and comfort to him when the going got tough.

FIGURE 5.1 Raúl A. Estrada-Oyuela in the midst of a huddle of climate negotiators. Photo courtesy of IISD/ENB/Leila Mead

The plot and protagonists

Estrada would certainly need every ounce of confidence, courage, and support that he could muster, if he was to reconcile the wildly differing concerns, demands, and expectations of the 160 participating delegations.

The Kyoto Protocol negotiations were all about strengthening the commitments of Annex I Parties – the developed countries. Among this group, there were three main camps. The EU (then just 15 strong), negotiated as a bloc, spending interminable hours in coordination meetings. The EU saw itself as a leader, pushing for an ambitious 15% cut in emissions to be applied to each developed country. The other Annex I Parties were mostly sceptical, both of the high level of aspiration, and the use of a uniform target.

The non-EU developed countries formed the so called JUSSCANNZ group, after the combined acronyms of the participating parties – Japan, US, Switzerland, Canada, Australia, Norway, New Zealand. This was only a loose formation, but in general, the JUSSCANNZ countries placed more emphasis on the need for cost-effectiveness and flexibility, and broadly supported emissions trading. The US was the trailblazer here, based on its successful experience in the trading of sulphur dioxide emissions permits from industrial smokestacks. Its vision of the protocol was of a 'cap and trade' system, whereby countries could 'trade' their allotted emission permits, providing an incentive for over-achievement and helping to level out mitigation costs. The EU, however, was deeply uncomfortable with emissions trading, fearing it would damage the treaty's environmental integrity. The developing countries were even more suspicious, viewing emissions trading as akin to the monetisation of the atmosphere, and premised on an ethically dubious 'right to emit'.

The third group were the economies in transition (some candidates to join the EU), the Eastern European countries whose economies and greenhouse gas emissions were still reeling from the collapse of the Soviet Union. Russian emissions, for example, had slumped by around two-thirds since 1989. Under a system of emissions trading, these countries might be massive sellers of credits, a situation that had not escaped their notice, or indeed that of potential buyers, notably the US and Japan.

The Berlin Mandate had established in crystal clear terms that developing countries would not be asked to take on new commitments at this stage. However, for the US, this presented real problems. The Administration, under Democrat President Bill Clinton and environmentalist Vice-President Al Gore, was out of step with its domestic political and economic elites, and the powerful fossil fuel lobbies who exerted such influence over them. Just a few months before the Kyoto Conference, the US Senate unanimously voted through the so-called Byrd-Hagel Resolution, signalling intent to veto any outcome that did not introduce 'new, specific, scheduled commitments for Developing Country Parties within the same compliance period'.[11] The US was in a quandary: how could it satisfy both the Berlin Mandate and Byrd Hagel? Wrestling with this impossible predicament was to become one of the dominant storylines of the Kyoto negotiations.

Resisting the pressure to assume new commitments unified the G-77. This unity, however, masked deep under-currents of division. Having failed to stop the adoption of the UNFCCC and the launch of negotiations in Berlin, large oil exporters – notably Saudi Arabia and Kuwait – now focussed their considerable resources on two main lines of fire: procedural delaying tactics, and financial compensation demands for any potential economic losses that might arise from a new protocol. It was common knowledge that these countries were backed by powerful US-based lobbying groups, notably the deceptively innocuous-sounding Climate Council and its shrewd chief lawyer, the late Donald H. Pearlman (dubbed the 'high-priest of the carbon club' by German magazine *Der Spiegel*).[12] All the way to the Kyoto conference, Saudi Arabia, Kuwait, and allies in the Organisation of Petroleum Exporting Countries (OPEC) pursued a barely concealed agenda of obstruction, aimed at achieving as little as possible, as late as possible.

This was in dazzling contrast to the Alliance of Small Island States (AOSIS), whose members – facing an existential threat from extreme weather and sea level rise – were desperate to achieve as much as possible, as soon as possible. Assisted by Northern environmental lawyers and NGOs, they had argued for Annex I Parties to commit to 20% emission cuts already in Berlin. Among most other developing countries, awareness of climate change and its implications was still rather hazy; the issue was largely seen as a 'first world problem' of uncertain relevance to them, except possibly as a new source of overseas finance. China was not yet the geopolitical and climate change giant it would become. However, along with Brazil, India, and other economic 'tigers' as they were then called, China was already aware that it would be next in line for demands to curb its emissions, something it would vigorously oppose as long as possible.

A sense of mission

It was Chairman Estrada's job to find a zone of agreement – however narrow it might be – among these protagonists and their often-conflicting priorities. He brought to this job a profound sense of mission. For Estrada, chairing the Kyoto Protocol negotiations was not just another diplomatic appointment. He passionately believed in the dangers posed by climate change and was determined not only that an agreement would be reached in Kyoto, but that it would be ambitious and meaningful. As Annex I Parties tabled their proposed emission targets throughout 1996 and 1997 – some much later than others – Estrada repeatedly urged them to raise the bar, especially those whose offers seemed weak. He was livid when the US originally pledged a mere stabilisation of emissions at 1990 levels, doing no more than freezing existing Annex I Party obligations under the UNFCCC. Estrada was similarly adamant that governments should be held accountable for meeting the targets they would pledge. He therefore strongly discouraged any talk of deadlines beyond 2010, believing such distant end points to be too easy for contemporary politicians to ignore. When he saw that consensus was crystallising around a 2008–2012 commitment period, he insisted on a clause requiring countries to have made 'demonstrable progress' by 2005.

While recognising the need for some flexibility, he was also clear that the treaty should not be riddled with loopholes. He was sceptical of emissions trading, and refused outright to seriously entertain proposals to allow the 'borrowing' of emission credits from future periods, denouncing these as 'escape routes'.[13] His reaction to increasingly convoluted proposed provisions on how to account for carbon absorption by forests and other so-called carbon 'sinks' was similarly blunt: 'We need to decide on a transparent system for everyone to understand. Stop playing with net-net, gross-net [different accounting methods], etc. We should not disguise what we're doing by picking and choosing sinks',[14] he chided. Although Estrada's efforts were not always successful – provisions on carbon sinks, in particular, remained murky – the Kyoto Protocol is certainly more rigorous and transparent because of his insistence on integrity.

Challenging obstruction

But not all countries were equally motivated. Throughout the two-year negotiations, Chairman Estrada had to fend off a barrage of procedural blocking tactics. The AGBM had barely got underway before OPEC nations insisted on being given a dedicated seat on the AGBM Bureau, an influential committee. The situation was resolved through a combination of informal assurances and stern words from Estrada:

> In the work of this Group, from the very beginning, we have had a group of delegations which I feel has tried to slow down the progress of our work. I shall do everything in my power to avoid that our feet get caught in a net of procedural matters.[15]

His resolve was tested again soon after, when OPEC members sought to interpret a legal clause (the 'six-month rule') in a way that would require negotiations to conclude a full six months before the treaty's scheduled adoption, a clearly unrealistic proposition. The environmental NGOs reported at the time:

> Defying both legal precedence and common sense, the forces of darkness attempted to use yet another procedural loophole to strangle any attempt at progress in the AGBM. With predictable cynicism, business lobbyist lawyers are apparently advising their OPEC clients that they can ambush the Berlin Mandate negotiations at the last minute by relying on Article 17.2 of the Convention [setting out the six-month rule].

(ECO, 1996)

This procedural ambush was again skilfully averted by Estrada. Responding with his lawyer's instincts, he sought an opinion on the matter from the UN Office of Legal Affairs, which quickly dismissed the strict interpretation in favour of a more workable version.

The most challenging reality that Estrada faced was the absence of any rules of procedure allowing adoption of a protocol by a vote; COP-1 had been unable to reach agreement on such voting rules, largely due to OPEC obstruction. This meant that any draft protocol would need to be adopted by consensus, but without any formal definition of the term. Consensus was not the same as unanimity. That much was clear. But beyond that, the legal waters were murky, and responsibility for interpreting and declaring consensus lay with the Chair. There was every prospect that, at the last minute, a small handful of parties – probably the oil exporters – might raise objections to the protocol presented for adoption. It would then be up to Estrada, as Chair, to determine whether there was still sufficient support – consensus – in the room to adopt the final document. Making the wrong call would be disastrous.

Estrada gave considerable thought as to how to avoid this eventuality. In an email to the secretariat in early 1997, he wrote:

> I spend hours thinking on how to avoid a blockage of our negotiations… The 'compensation' idea [an OPEC proposal whereby they would receive financial compensation for any loss of oil revenue resulting from climate policies] was introduced to create a package and that package plus the need for consensus are an excellent combination to block the process.[16]

His creative solution was to highlight the possibility of crafting the substance of the agreement not as a protocol, but as an amendment to the Convention; this was because the Convention itself stated that amendments could be passed by just a three-quarters majority vote. Although all the negotiating texts issued in 1997 were structured in the form of a protocol, Estrada underscored that their contents could be redrafted in the form of an amendment, enabling adoption by a vote. In the weeks leading up to Kyoto, he instructed the secretariat to do just that. Throughout the Kyoto Conference, Estrada therefore had a draft amendment text in his back pocket as a contingency, just in case. In the end, it wasn't needed, but Estrada thought ahead and took no chances.

While Estrada had made it clear throughout the negotiation process that he would not allow the emerging treaty to be held hostage by a few laggards, he upped the stakes as the Kyoto Conference approached. Most dramatically, he caused near panic among delegates and his UN secretariat team when, at AGBM 8 just before Kyoto, he called for a vote on a piece of rather inconsequential text, to which Australia, Canada, and the US were objecting. He did so on the somewhat shaky (but defensible) legal grounds that those three countries were challenging his Chair's ruling that consensus existed on the text (procedural votes challenging a Chair's ruling are permitted). Many delegates pleaded with him to back down, fearing that a precedent would be set. An unruffled Estrada eventually desisted, but not before declaring that the challenge to his ruling had been withdrawn, and the consensus stood. The incident ended with a boost to Estrada's authority. Any delegation who might have thought to block consensus in Kyoto would now think twice.

Unafraid of confrontation

Chairman Estrada did not hesitate to confront delegates whom he deemed to be out of line. His fire was directed mostly at the Annex I Parties, repeatedly chiding them for weak leadership, failure to cobble together a common position on emissions trading, and slowness to put their cards on the table for genuine negotiation. He was particularly critical of the EU, whose constant internal coordination meetings irritated him. In his memoirs, G-77 Chairman Mwandosya describes 'a not-so-diplomatic exchange of words between Chairman Estrada and the Right Honourable John Prescott [UK Deputy Prime Minister] who complained that Chairman Estrada had shown insensitivity to, and lack of appreciation of, the role of the European Union' (p. 122). But ever the diplomat, Estrada typically maintained excellent personal relationships with his sparring partners outside the negotiation chamber. He fondly recalls reminiscing with Prescott over their Kyoto days when their paths crossed again at the Rio+10 Conference in Johannesburg in 2002.[17]

As a developing country delegate, it was politically acceptable for Estrada to challenge his fellow G-77 members in a way that would have been impossible for a developed country Chair. He certainly did not shy away from doing so. 'He was rude to everyone', chuckled a Ghanaian delegate. But as the deadline approached and the pressure mounted, Estrada started to lose his temper, pushing the boundaries of diplomacy. In the dying days of Kyoto, he provoked a walk-out from the Brazilian chief negotiator when he implied, in plenary, that Brazil was seeking financial gain from its proposed clean development mechanism. He accused the Russian Federation of similar acquisitiveness over emissions trading, and publicly remonstrated with a senior Mexican delegate for apparently messing up a chairing task he'd been assigned. Although he sometimes went too far in his aggression, that Estrada did not court popularity was indispensable. Jennifer Morgan, veteran environmental activist and now Executive Director of Greenpeace International, reflected on this:

> I think having someone with his personality and his initiative to kind of take control of situations, to be unpopular at times, and not to please absolutely everyone 100% of the time, is necessary in a strong Chairman, otherwise you'll never get an agreement.

Ian Fry, negotiator for Tuvalu, echoed this sentiment: 'He was very strident in his approach… And forceful and [he] offended some delegations in doing that. But I don't think we would have got an outcome otherwise'.

The consummate chair

For Zammit Cutajar, one of Chairman Estrada's great strengths was 'the sheer ability to chair a big meeting I have never seen before or since'. Estrada was always in complete control of proceedings: not for him the chaos and confusion of subsequent infamous COP meetings.[18] In the words of a US State Department memo,[19] Estrada was 'a showman with a zest for drama', who positively relished presiding over heated debates, vigorously wielding his gavel to rein in unruly delegates. Significantly, Estrada had an uncanny instinct for reading 'the sense of the room' and acting accordingly. He knew when to make his move with a landmark proposal, and when to back off and wait. He knew when to let delegates talk themselves out, and when to cut them short, something he did with increasing frequency as the negotiations approached their finale. A Thai delegate recalled fondly: 'With Ambassador Estrada, no one could speak nonsense for too long. He cut it!'. Estrada understood that intergovernmental negotiations unfold through an interplay of both profound geopolitical forces and the idiosyncrasies of the individuals involved. For him, the most important meeting document was the list of participants, which helped him to call on delegates by name. He had his eye on the more difficult characters in the room – Mohamed Al Sabban from Saudi Arabia, the late Bernardita de Castro Muller from the Philippines, amongst many others – and was ready to parry their anticipated objections with humour or sheer force of personality. For a senior UK lawyer, '...this priceless ability to keep people's humour high, preventing irritations spilling out into anger... actually in negotiations, it's far more important than a detailed knowledge of the subject'.

Political awareness and issue management

Although it is true that Chairman Estrada did not have detailed technical knowledge, he had a deep sense of the political dynamics at work and how the different issues on the table should be managed. This made him 'a political and tactical force to be reckoned with', according to Richard Kinley, his main collaborator within the UN secretariat. Farhana Yamin, a UK-based lawyer supporting AOSIS nations, commented,

> I think he understood the bottom line of all the positions of the Parties, which is what his job as the Chairman is really about...he was a very great adept at identifying what everyone's interests were and trying to work within those.

Although political instinct played a part, such deep understanding was also the product of practical thinking and hard work. In mid-1997, Estrada asked the UN secretariat to prepare all the written proposals submitted by governments into a single, cross-referenced, thematically organised compilation. This well-thumbed document became a constant source of reference for Estrada as he sought to devise a political package acceptable to all. Crucially, this meant that his Chairman's drafts were all derived from original proposals of the parties – bottom up, in today's parlance. A good example of Estrada's judicious issue management concerns one of the Kyoto Protocol's most innovative elements, the clean development mechanism (CDM). Under the CDM, developed countries can invest in projects that curb emissions in developing countries, and use some of the resulting credits to count against their own targets. Originally proposed by Brazil as a 'clean development fund' into which non-compliant Annex I Parties would pay, this pioneering mechanism was worked on behind-the-scenes by its Brazilian brainchild, Luiz Gylvan Meira Filho, US negotiator Dan Reifsnyder, and eventually also the late John Ashe. Ashe, a wily delegate from Antigua and Barbuda, managed to slip in a ground-breaking new channel for adaptation financing, thus securing the support of climate vulnerable developing countries. Estrada didn't like this kind of offsetting mechanism, fearing it could become a loophole. But he saw how politically useful it could be, both to the US as a demonstration of developing country action, and to those developing countries as a source of much-needed finance. He also understood that, if the emerging text on the CDM were exposed to full scrutiny too early, it would be suffocated at birth. The CDM was so new, so potentially controversial, yet so politically promising as a mechanism bridging North and South, that it had to be worked on – not in secret exactly, but certainly under a 'veil of uncertainty' (Oberthur and Ott, 1999, p. 167). Estrada thus let the talks on the nascent CDM run their course behind-the-scenes, while making sure he was kept abreast of progress, often over dinner with Meira Filho and both their wives. His patience and trust in those three experienced diplomats – Meira Filho, Reifsnyder, and Ashe (arguably heroes in themselves) – paid off. The CDM was critical to forging a narrow zone of agreement for parties to land on.

Not infallible

Estrada was an excellent and committed Chair. But he was not a magician, and his efforts did not always bear fruit. In an attempt to find a compromise acceptable to both the US and developing countries, Estrada prepared a draft article setting out a process whereby developing countries could take on voluntary obligations under the protocol, as and when they saw fit. Although the draft was loosely based on an AOSIS proposal, it was not well-received; in retrospect, it was perhaps too elaborate, raising too many questions that scared off potential supporters. Estrada's suggestion to inscribe emission targets not in an annex, which could only be amended through a cumbersome procedure, but in a less formal 'attachment' that could be more easily updated, was also dismissed.[20]

Although Estrada can be credited for raising the bar of emission targets, not all governments proved susceptible to his cajoling. When pressed by Estrada to consider a stronger target than the derisory 0% his country had put on the table, the

Russian delegate declared 'not now, not ever' (ENB, 1997). Such intransigence was fuelled by Russia's resentment at suggestions – including from Estrada – that it sought to profit financially from emissions trading, while the delegation felt neglected by the diplomatic and political attention that was revolving primarily around the US, EU, and Japan.

Australia similarly remained impervious to any encouragement to raise its paltry ambition, even managing to slip in a last-minute clause on forestry accounting – diluting its already weak target – in the dying minutes of the conference, when most delegations were too exhausted to notice or care. Always on top of things, Estrada had of course noticed and cared, but deemed the concession 'less bad than leaving it [Australia] with no percentage at all'.[21] Australia certainly extracted full advantage from a diplomatic faux pas, whereby Estrada stated to the media early during COP-3 that Australia could not join the Kyoto Protocol with the pathetically weak target it had proposed. With Australia's chief climate official, Meg MacDonald, feigning great offence, Estrada was forced to retract and seemingly lost any leverage to push that great recalcitrant to a more reasonable climate commitment.

Time management

Negotiations tend to be remembered for their dramatic finales unfolding in the last 24 hours of the culminating conference. But the seeds of success (or failure) are typically sown throughout the process, in the preceding months or years. One of the great dangers of large-scale multilateral negotiations is that they get bogged down in their own complexities, so that delegates fail to grasp a potential compromise that might be within their reach. If a negotiation finale is confronted with a mammoth negotiating text hundreds of pages long, then even with the best will in the world, delegates will struggle to finalise a consensus outcome. If the inherent tendency of governments to procrastinate is indulged, and every little issue is left to the last minute, then time will literally run out or exhaustion will overwhelm the ability of delegates to make sensible decisions. As an experienced negotiator and Chair, Estrada knew this full well, and worked hard to build momentum in the process. The first 18 months of negotiations were devoted to an 'analysis and assessment' of options; this was not only an important intellectual and trust-building exercise, but it also served a strategic purpose. Estrada knew that it would be impossible for the US delegation to make any kind of serious proposal in an election year. With President Bill Clinton re-elected in November 1996, the US position was clear, and Estrada could crank the process up a gear. Sure enough, at AGBM 6 in March 1997, he declared that the time was now ripe for negotiation and started to interrupt any delegate seeking to just restate well-worn positions. Estrada insisted that delegates begin work on a draft negotiating text already at AGBM 7 in July, produced his own Chairman's text at AGBM 8 in October, and from then on refused to reintroduce any obviously unviable proposals or new square brackets (signalling areas of disagreement) into the emerging document. Although some delegates protested vigorously, Estrada stood firm. At the same time, he closed all informal groups to observers, raising the ire of NGOs, but sending out a clear signal that it was time for government delegates to get serious.

The result was that, by the start of the Kyoto Conference, delegates were ready for hard bargaining, based on a perfectly readable 32-page draft of the emerging protocol. There was already a semblance of provisional agreement on some of the less controversial clauses, allowing delegates to focus their time and attention on the 'crunch issues'. Estrada sought to raise the tempo of talks throughout COP-3. At one point, he locked a group of ministers in a room, threatening to keep them there without food or drink, until they reached agreement on emissions trading and other political hot potatoes (his plan was thwarted when the EU called up the venue caterers, and a trolley of lavish refreshments was wheeled in). He pushed the Annex I Parties hard to finalise their emission targets, knowing the necessary horse-trading among them would take time. Frustrated by the lack of progress, he eventually circulated his own draft list of country-by-country emission reduction targets, with figures deliberately inflated by a few percentage points, at a hastily convened high-level luncheon meeting. The reaction of delegates was, as Kinley recalls, 'like a bomb going off... Everyone suddenly lost their appetite. Eizenstat [chief US negotiator] even declined a salad'.[22] A few hours later, officials were queuing up at Estrada's office door as if at a confessional, to reveal what they could, and could not, accept. He released deliberately provocative text also on other issues – such as carbon sinks, or the entry into force threshold of the protocol – which similarly galvanised (or scared) governments into putting their cards on the table. Many other examples could be given of how Estrada strategically managed the two-and-a-half-year negotiation process. While unglamorous, they were critical to the triumph of the final night.

The finale

And indeed, Estrada will always be remembered for his masterful chairing of that historic last night in Kyoto. Typically, the endgame in such large multilateral negotiations is hammered out behind closed doors, among a small group that may or may not represent the full complement of participants, before being presented as a package deal to a plenary meeting for formal adoption. Not so in Kyoto. Estrada insisted that the finale should take place in an open room – in the presence of NGOs, the media, and even a live internet webcast (pioneering technology at the time) – so that, should the talks collapse, everyone would know who was responsible. He also requested that the meeting take place in the smaller of two plenary rooms, to create a more intimate atmosphere. In the end, the room was so packed, 'negotiations would have been possible without a microphone... delegates felt almost physically the pressure to arrive at a deal' (Oberthur and Ott, 1999, p. 88). The result was a uniquely transparent and open negotiation finale, which became 'one of the most exciting nights of international environmental diplomacy' (Oberthür and Ott, 1999, p. 88). Interviewed for the 20th anniversary of that night, Kinley recalled with emotion how it was 'all done in public... This level of transparency and openness in finalizing a legally binding document has never been repeated since in the climate process... it was an amazing experience'.[23]

The start of the meeting was postponed several times, as EU and JUSS-CANNZ countries made a last-ditch attempt to bridge their differences on emissions trading, the G-77 tried to coordinate the divergent views of its members on several issues, and the secretariat prepared the final draft protocol that delegates would work on. When that document finally rolled off the press in the early hours of Thursday, 11 December, UN security guards had to escort the distribution team to the plenary room, lest they be mobbed by delegates desperate to get their hands on a copy. And where was Estrada while all this was going on? He was in his hotel, resting and refuelling. As he later explained to a journalist from the Asia Times,

> To conduct the negotiation, you have to be very relaxed and awake ... I went to my hotel...I had dinner with my wife, and I slept for three hours while the delegates were fighting about the points. Then I returned fresh, and they were tired.
>
> *(Hurst, 2017)*

Estrada had planned the endgame meticulously. Thanks in large part to his judicious issue and time management, many elements of the draft protocol had already been agreed, at least in principle: the baseline for targets (1990), approach and timing of the deadline (a 2008–2012 commitment period), coverage (all six main greenhouse gases as a basket, plus some forestry activities), provisions for reporting and review, entry into force thresholds, and many more. But the three main pieces of the political puzzle were outstanding: emissions trading, how developing countries might participate, and the level of the emission targets themselves.

Estrada opened up the draft text for comment, clause by clause. He made it clear that delegates could suggest changes, but if there was no consensus to make a change, then the existing text would stand. If anyone challenged his ruling, he would take it to the vote, as he had done at AGBM 8. This worked well for the first few draft clauses and a sense of optimism and momentum began to spread across the room. But there were flashpoints ahead. At about 3:30 am, negotiators came to draft Article 3, paragraph 10, on emissions trading. This time, Estrada knew that he had to allow an exhaustive exchange of views. More than 50 delegations spoke up, including some who had never taken the floor throughout the whole two-and-a-half-year process. India's Vijay Sharma spoke with great passion about the inequitable commercialisation of the atmosphere. Most other developing countries agreed, suggesting instead further study of this new concept, and the EU was lukewarm at best. But the US needed emissions trading. Now.

With remarkable candour and emotion, chief negotiator Stu Eizenstat pleaded with developing countries to let the paragraphs go through. This marked the low point of the whole negotiation process. Estrada responded that he was 'really concerned' that this issue 'could blow up the whole protocol' (UNFCCC, 2000, p. 392). With that, he suspended the meeting. About 45 minutes later, having met in a side room with Eizenstat, UK Deputy Prime Minister Prescott, and Zammit Cutajar, he returned. The room was deathly silent as Estrada presented a new, finely balanced compromise. Three short sentences would now be located right at the back of the protocol (away from the article on emission commitments). They would include the bare essentials needed to activate emissions trading, making it clear that this new mechanism would not replace domestic action, and would be subject to rules to be decided in another negotiating round. Although India and others began to ask for the floor, Estrada declared there was consensus, brought down his gavel, and moved onto the next clause. India did not protest. Estrada's gamble had paid off. Axel Michaelowa, a German researcher, later wondered

> What I am really intrigued [about] is that nobody dared to put up the flag [asking for the floor to object] after he hammered through … How he sensed how far he could go, and where he should stop. Only a real diplomat could grasp this, and he did.

But the work was not yet done. Another piece of the political puzzle had to be skilfully managed: provisions for developing countries. Here, there was a delicate balancing act to perform between the US imperative that there should be something in the Protocol that it could sell back home as developing country action, and the implacable opposition of the G-77 to anything going even a centimetre beyond the Berlin Mandate deal of 'no new commitments'. The document, as it stood, included the article on 'voluntary commitments' that had been drafted by Estrada himself, as a compromise that might, just might, meet both camps halfway. Estrada once again took the time, despite the lateness of the hour, to listen to over 25 interventions. Opinions were split: some developing countries, notably AOSIS, saw the merits of the draft article, but China, India, and other emerging economies (who feared pressure to curb their emissions, including from the development banks) were unbending. Several delegations, including Argentina, Mexico, and South Korea, started to propose complicated textual amendments, which were sowing confusion. The talks were once more on the brink, and Estrada knew he had to pull them back. With the Mexican delegate still in full flow, Estrada declared it was 'too late' to consider such amendments, there was no consensus on the draft article, and it would be deleted. The room held its breath. Would US delegates object? In a press conference a few hours later, Eizenstat would decry deletion of this draft article as 'our single greatest disappointment'.[24] But he did not object. On this most potent of issues, Estrada had taken the developing country side – a trade-off for having conceded emissions trading to the US – and the gamble had paid off. The late Kok Kee Chow, an experienced

Chairman from Malaysia who understood the pressures of the role, recognised Estrada's bravery: 'I really salute him ... he made a very bold decision'.

The core political agreement had been struck. The package deal – crucially including the CDM that was gavelled through a couple of hours later – would fly. As the negotiations moved on to the more straightforward final clauses of the treaty, Estrada sent Kinley down from the podium to discreetly check with some delegations that the emission targets they had tentatively pledged were still holding. They were. Finally, and with a minimum of fanfare, the heads of Annex I Party delegations approached the podium to individually confirm their targets, through handwritten notes on scraps of paper. Once they were all received, Estrada announced to the packed plenary room that these added up to a collective 5% reduction from 1990 levels – not as ambitious as he had hoped, but nonetheless a 'significant departure from historical emission trends', as Zammit Cutajar later reported to the UN Secretary General.[25] The final piece of the puzzle was in place. In the end, Estrada did not have to invoke his contingency plans. Even the OPEC countries, whose threat of last-minute blocking had cast such a shadow over the whole process, were remarkably acquiescent on that last night.[26] As Chairman Estrada brought down his gavel on the final text, the exuberant scenes that ensued were more reminiscent of a major sporting event than a UN conference. At that moment, on that last night in Kyoto, it seemed like the world's governments had finally risen to the challenge. There was hope that the looming threat of dangerous climate change might be averted.

FIGURE 5.2 Estrada in charge as he chairs a meeting. Photo courtesy of IISD/ENB/ Leila Mead

United States rejection

But the euphoria was short-lived. Hours after its adoption, US Senators and industry lobbyists loudly denounced the deal, claiming that the Kyoto Protocol would wreak havoc on the US economy while letting China and Mexico off the hook. The Kyoto Protocol was eventually signed by the Clinton Administration, but never sent to the US Senate for its 'advice and consent', which would be needed for ratification. Right-wing organisations and fossil-fuel lobby groups funded a multi-million-dollar misinformation campaign ('fake news' in today's words) about the supposed dire consequences of the Protocol for the American way of life. Even US academics lined up to criticise a caricatured version of the treaty. 'Kyoto' soon became a dirty word in the US. And this despite the treaty including virtually every US negotiating demand. Taking office in January 2001, Republican President George W. Bush soon declared his opposition to the Kyoto Protocol 'because it exempts 80 percent of the world, including … China and India… and would cause serious harm to the US economy'.[27] In his memoirs, Swedish Ambassador Bo Kjellén – who played a central role in the negotiations not only on the Kyoto Protocol, but also the UNFCCC and Berlin Mandate – remembers how US national security adviser Condoleeza Rice told an aghast group of EU Ambassadors over lunch in Washington that the Kyoto Protocol 'was dead' (Kjellén, 2004, p. 97). Such disregard for an international treaty initially had the opposite effect, rallying the rest of the world in support of the Kyoto Protocol. Its implementation rule book was eventually hammered out, and the treaty finally came into force on 17 February 2005, following the belated ratification of the Russian Federation (many say enabled by EU promises to support that country's entry into the World Trade Organisation).

The Kyoto Protocol was extended for a second commitment period from 2013 to 2020. This was a major political achievement, but by now, the absence of the US had taken its toll, and several other large economies refused to join, including Japan. Countries with emission targets under the Kyoto Protocol's second commitment period (known as the Doha Amendment) thus accounted for only about 13% of global emissions. In a twist of fate, the Doha Amendment – whose entry into force required ratification by three-quarters of the Kyoto Protocol's parties (144) – only formally came into effect on 31 December 2020, the very last day of its application. With the 2015 Paris Agreement now providing a new framework for action for the post-2020 period, the Kyoto Protocol – while not 'dead' – has, to all intents and purposes, run its course.

The failure of the US to ratify the Kyoto Protocol did not kill the treaty. But without the world's largest historical emitter, political superpower, and leading economy, the Protocol's impact was seriously weakened. The Protocol was always intended as a long-term instrument that would gradually expand to new entrants, but absent US participation, there was zero chance of persuading the large emerging developing country emitters to join the Kyoto club with their own binding emission targets. And so, the emissions of emerging economies rose, and with them, the global emissions curve. Had the US stayed with the Kyoto Protocol, met its target, and led the world to a second, stronger, commitment period, it is possible to imagine that developing countries – including China – might have followed, and started to curb their emissions much earlier. Perhaps history would have been different, had climate-sceptic George W. Bush not defeated climate-enthusiast Al Gore in a hotly contested US Presidential election in 2000 marred by alleged electoral irregularities. It is chilling to think that the fate of the world's climate may have been decided by a few thousand disputed ballots and outdated voting technology in the US state of Florida.

An unsung legacy

Even without US support, however, the Kyoto Protocol marked a critical and enduring step forwards towards the ultimate objective of avoiding dangerous climate change. All the parties with emission targets met their goals, under both commitment periods (a handful using emissions trading). As a group, the countries with targets reduced their emissions 22% from 1990 levels over the first commitment period, and by more than 28% for the second period (2018 figures),[28] vastly exceeding both collective targets.

With its legal obligation to fulfil binding targets, the Kyoto Protocol prompted the rollout of emission-curbing legislation among advanced economies. But the legislative impact extended further, as low-carbon norms and policy experiments began to diffuse more widely beyond the Kyoto club. The number of climate laws in major economies jumped from under 40 in 1997, to nearly 500 at the end of 2013 (Fankhauser et al., 2016). Researchers estimate that, between 1999 and 2016, about 37.7 Gt of CO_2 (about a year's worth at current rates) were saved by climate laws, the bulk from Kyoto Protocol parties (Dubash, 2020; Eskander and Fankhauser, 2020).

One of the ground-breaking features of the Kyoto Protocol was its introduction of market mechanisms, through emissions trading and the CDM.[29] These were pioneering policy instruments, which were to spread across the globe. In a quirk of dramatic irony, some of their most vociferous critics in Kyoto – the EU and China – became the first to embrace emission trading schemes.

The impact of the CDM has been particularly far-reaching, generating over 2 billion emission credits (each representing a ton of CO_2) from projects ranging from waste management, to wide-scale forest regeneration, to the replacement of highly polluting biomass cookstoves with clean solar powered ones. In India alone, 45 million high-efficiency light bulbs were distributed to households through the CDM (Malhotra et al., 2021). Co-benefits from the over $300 billion invested have included job creation, technology transfer, and improved air quality (UN CDM, 2018). The CDM thus became pivotal in changing the conversation about climate change mitigation in developing countries: projects encouraged experimental learning, helping to gradually build up confidence in the decarbonisation agenda, a shift in attitude that would help bring about the universal pledging approach of the Paris Agreement. The Adaptation Fund created by the CDM was the climate change regime's first dedicated adaptation financing channel. While its funds are still tiny compared to the scale of the problem – over $200 million raised over the Protocol's lifetime – it is of immense importance to developing countries, who have insisted that it should also serve the Paris Agreement.

But perhaps the key legacy of the Kyoto Protocol was political. First, the Kyoto Protocol helped to boost, widen, and sustain awareness of the climate change problem, ensuring that even its detractors could no longer ignore the issue. The US was stung by the opprobrium of the international community when it rejected the Kyoto Protocol – as was Canada, when it withdrew from the treaty in 2012 – and pressure on the laggards to take action only grew over time. Businesses, local governments, financial institutions, development banks, and other stakeholders also took notice. A decarbonisation norm began – slowly but surely – to take hold. Yamide Dagnet, director of climate at the World Resources Institute, agrees: 'the Kyoto Protocol played a critical role by driving innovation, raising awareness and catalysing climate action'.[30]

Second, the Kyoto Protocol began to fulfil the 'grand bargain' of the UN-FCCC, that is, for the developed countries 'to take the lead' in addressing climate change. Without the (admittedly weak) leadership embodied in the Kyoto Protocol and its Doha Amendment, developing countries would never have even considered launching the negotiations that led to the Paris Agreement and its universal pledges. Legal experts agree. According to the official history of the Paris Agreement, it was the entry into force of the Kyoto Protocol, followed by the adoption of the Doha Amendment, that unlocked negotiations on the future regime (Bulmer et al., 2017, p. 55).

Richard Kinley – who continued to work at the UN Climate Change Secretariat for 20 more years, playing a key role in supporting the negotiations on the Paris Agreement – summed it up neatly:

> The Kyoto Protocol meant that developed countries (most of them at least) could say they had fulfilled their leadership obligation. And I hate to think where the world would be if it had not had the Kyoto Protocol to rally round during some pretty dark days – that may well be its enduring legacy.[31]

Back to our hero

And what of Ambassador Estrada? With his posting in Beijing coming to an end, our protagonist returned to Buenos Aires. Inexplicably, his political masters greeted him with indifference. He was reduced to attending COP-4 taking place in Buenos Aires as a driver for the Panamanian delegation until Zammit Cutajar, horrified at this humiliation, authorised a special guest badge. By contrast, Estrada found himself in great demand from US Ivy League universities.

He became a visiting professor, regaling international relations students with dramatic tales from the negotiations. Following a change in government, he returned to the Argentinian Foreign Ministry in 1999, serving for a period as Chair of the UNFCCC's Subsidiary Body for Implementation and the Kyoto Protocol's Compliance Committee. But he never quite recaptured the personal authority and political alchemy that he had wielded to such good effect in the Kyoto Protocol negotiations. Relations with the Argentinian government remained tense. In September 2007, Estrada – now Special Representative for International Environmental Negotiations – said in public that Argentina had 'no environmental policy'.[32] It was the final straw. Estrada's high reputation meant he could not be sacked, and so his post was simply abolished. His career as an international diplomat was over.

In between spending time with his many children, grandchildren, and great-grandchildren, Estrada – now well into his 80s and mourning the loss of his wife – is still exploring practical means of boosting climate action, with a focus on engaging cities, businesses, and other stakeholders. 'We need to look for the means to turn mitigation into good business', he tells me in a recent email exchange. And as the climate negotiations face new challenges, he itches to get back in the saddle: 'I would like to start over again', Estrada confides. 'But I have no chance to do so', he adds wistfully.

Twenty-five years after that historic last night in Kyoto, Ambassador Raúl A. Estrada-Oyuela has earned his place in the history books of environmental diplomacy. In the end, the true promise of the Kyoto Protocol was thwarted by the deeper political and economic forces that continue to slow down the international response to climate change. But Estrada showed the world what could be achieved with political will, imagination, and courage. And for that, he will always be rightly known as the 'hero of Kyoto'.

Notes

1 Remarks on the occasion of Ambassador Estrada's receipt of the 2006 International Environmental Law Award from the Center for International Environmental Law (CIEL). Available online at: https://www.ciel.org/about-us/2006-international-environmental-law-award-recipient-raul-estrada-oyuela/

2 Memorandum from Michael Zammit Cutajar to Nitin Desai, Under-Secretary-General of the UN Department for Economic and Social Affairs, dated 12 December 1997. On file with author and UN Climate Secretariat.

3 The negotiations that culminated in the UNFCCC have their own stories of high drama and heroism to tell, not least the courage of French Chairman Jean Ripert, who held his nerve and brought down his hammer to adopt the treaty, despite oil exporting countries waving their nameplates in the air to object. Without this act of wise judgement and bravery, the international response to climate change might have been stillborn.

4 Originally 36, now 41. Changes have been made to reflect new European entrants and changes to national boundaries.

5 Emissions data. Available online at: https://edgar.jrc.ec.europa.eu/overview.php?v=CO2andGHG1970-2016&sort=des2, GDP data from UNDP, 1998.

6 Indeed the two-way division was never as rigid as often assumed, with both least developed countries and economies in transition enjoying special treatment.

7 AGBM 1 opening plenary, 21 August 1995. Cassette recording on file with the UN Climate Change secretariat. All further references to statements made during negotiations are also on file with the UN Climate Change secretariat.

8 Zammit Cutajar went on to recall how, at the close of COP-1 in Berlin, environmental activists had invaded the podium to protest that no new treaty had been adopted. Estrada jumped to COP President Angela Merkel's defence, throwing the protesters off the stage before UN security could rush to the scene.

9 Unless otherwise stated, all quotations from delegates are taken from interviews conducted by the author in 1999, which are on file with the author.

10 Dated 12 March 1997. On file with UN Climate Change secretariat.

11 S.RES 98. *Expressing the sense of the Senate regarding the conditions for the United States becoming a signatory to any international agreement on greenhouse gas emissions under the United Nations Framework Convention on Climate Change.* July 25 1997.

12 See Hohenpriester im Kohlenstoff-Klub, Der Spiegel 14/1995, 02.04.1995. Available online at https://www.spiegel.de/politik/hohepriester-im-kohlenstoff-klub-a-2cf26de6-0002-0001-0000-000009180353?context=issue

13 Fifth plenary meeting of the Committee of the Whole, 4 December 1997.

14 Eleventh plenary meeting of the Committee of the Whole, 8 December 1997.

15 Opening statement to AGBM 3, 5 March 1996.

16 Dated 13 March 1997. On file with UN Climate secretariat.

17 Email exchange, dated 3 October 2021.

18 These famously include the final plenary of the 2009 Copenhagen Conference, discussed by Chris Spence in the next chapter 'The Copenhagen Climate Summit: Barack Obama – the missing hero?'

19 Approach to Estrada (aka 'Estrategy'). Unclassified Department of State. Doc no. C05577166. Dated 3/18/97.

20 In this case, it seems that Estrada was ahead of his time. One of the greatest criticisms of the Kyoto Protocol was that it was 'inflexible' in its listing of parties, yet Estrada did everything he could to introduce greater flexibility – it was governments who resisted this. The second commitment period to the Kyoto Protocol, adopted in 2012, introduced more flexible amendment procedures, and the Paris Agreement went even further, stating that Parties could adjust their existing emission pledges 'at any time' (Paris Agreement, Article 4.11).

21 Email exchange, dated 3 October 2021.

22 Interview transcript dated 1 December 2017, and personal email exchange, 29 September 2021, both on file with author.

23 Interview transcript dated 1 December 2017, on file with author.

24 Quoted in *Implications of the Kyoto Protocol on Climate Change. Hearing before the Committee on Foreign Relations.* United States Senate, 105–457. February 11, 1998.

25 Note for the Secretary-General. The Kyoto Protocol to the UNFCCC: Analysis and follow-up. On file with UN Climate Change secretariat.

26 Oberthur and Ott (1999) suggest, very plausibly, that the Japanese hosts offered trade incentives to Saudi Arabia in return for its cooperation in the negotiations. It may also be that Saudi Arabia and others were reassured by their oil lobbyist friends that the US would never ratify the Kyoto Protocol anyway.

27 *Letter to Members of the Senate on the Kyoto Protocol on Climate Change,* 13 March 2001, available at https://www.presidency.ucsb.edu/documents/letter-members-the-senate-the-kyoto-protocol-climate-change.

28 On the first commitment period, see https://unfccc.int/news/all-hands-on-deck-for-the-doha-amendment; on the second, see Annual compilation and accounting report for Annex B Parties under the Kyoto Protocol for 2020. Note by the secretariat (unfccc. int), Section B.1. https://unfccc.int/sites/default/files/resource/cmp2020_05E.pdf

29 Neither emissions trading nor the CDM are without their critics, including Estrada. It is far beyond the scope of this chapter to engage in critical analysis of the detailed merits and demerits of these instruments. Any interested reader only has to type the words into a reputable internet search engine to begin their own investigation.
30 Quoted   in   https://www.climatechangenews.com/2020/10/02/nigeria-jamaica-bring-closure-kyoto-protocol-era-last-minute-dash/
31 Interview transcript dated 1 December 2017, on file with author.
32 Email exchange, 4 October 2021.

References

Dubash, N. (2020). Climate laws help reduce emissions. *Nature Climate Change*, 10(8), 709–710.

ECO (1996). *Good lawyers, bad lawyers*. Newsletter of the Climate Action Network. 12 July. On file with author.

ENB, 1997. *Report of the third conference of the parties to the United Nations framework convention on climate change*. 1–11 December 1997, Vol. 12, no.76, 13 December 1997. Available online at: https://enb.iisd.org/events/unfccc-cop-3/summary-report-1-11-december-1997

Eskander, S. M. & Fankhauser, S. (2020). Reduction in greenhouse gas emissions from national climate legislation. *Nature Climate Change*, 10(8), 750–756.

Fankhauser, S., Gennaioli, C. & Collins, M. (2016). Do international factors influence the passage of climate change legislation? *Climate Policy*, 16(3), 318–331, doi: 10.1080/14693062.2014.1000814

Hurst, D. (2017). *Kyoto Accord 20 years on: Hard won but inadequate*. 10 December 2017. Available online at: https://www.asiatimes.com/article/kyoto-accord-20-years-hard-won-inadequate/

Kjellén, B. (2014). *A new diplomacy for sustainable development: The challenge of global change*. Routledge.

Malhotra, A., Mathur, A., Diddi, S. & Sagar, A.D. (2021). Building institutional capacity for addressing climate and sustainable development goals: Achieving energy efficiency in India, *Climate Policy*, doi: 10.1080/14693062.2021.1984195

Mwandosya, M. (2000) *Survival emissions: A perspective from the South on global climate change negotiations*. Dar es Salaam: Dar es Salaam University Press and Centre for Energy, Environment, Science and Technology.

Oberthür, S. & Ott, H. (1999) *The Kyoto Protocol: International climate policy for the 21st century*. Berlin: Springer Verlag.

UN CDM (2018). *Achievements of the clean development mechanism: Harnessing incentive for climate action*. Available online at: https://unfccc.int/sites/default/files/resource/UNFCCC_CDM_report_2018.pdf

UNFCCC (2000). *Tracing the origins of the Kyoto Protocol: An article-by-article textual history*. FCCC/TP/2000/2. Available online at: https://unfccc.int/documents?f%5B0%5D=symbol%3Afccc/tp/2000/2

6

BARACK OBAMA

The Missing Hero? The Copenhagen Climate Summit

Chris Spence

It was not supposed to be like this. When President Barack Obama's plane touched down in Denmark on a windswept, wintry morning in December 2009, the Copenhagen Climate Conference was entering its last day. The 44th President of the United States was supposed to arrive just in time to sign and celebrate a major new climate agreement. This outcome, which would be endorsed by dozens of world leaders, would mark the end of two weeks' hard negotiations and months of painstaking preparations. For Obama, it was a chance to bask in the glow of the global media spotlight and celebrate a landmark environmental victory—a deal that would stop dangerous climate change in its tracks. After a year fighting a worldwide recession and facing stubborn Republican opposition on Capitol Hill to his healthcare reforms, the Danish trip should offer a welcome change in mood.

It was not to be. Instead, Obama arrived at the Bella Conference Centre in Copenhagen to a chaotic scene. Talks had stalled and were descending into chaos. Negotiators had failed to break down differences on dozens of seemingly intractable problems. Delegates were in disarray. Tempers were frayed to breaking point.

In an eleventh-hour attempt to save the day, their Danish hosts had proposed a new, simplified draft agreement. Rather than helping, however, this well-intentioned move had only made things worse. For a start, it infuriated many diplomats who felt it invalidated their earlier efforts. Second, it had been introduced with the foreknowledge of only a handful of insiders, blindsiding many delegations and leading to charges of underhandedness and dishonesty. Meanwhile, organizers had decided to limit the number of delegates allowed inside the conference centre once world leaders arrived, leaving thousands of conference goers literally out in the cold.

Obama's hopes of enjoying a well-crafted success story were in tatters. He was being swept into a perfect storm, with delegates drowning in complicated and disputed documents and apparently no way to avoid disaster. It was a media

DOI: 10.4324/9781003202745-7

debacle in the making—and a catastrophe for our planet. If seasoned delegates steeped in the minutiae of one of the most complex global treaties ever had failed after two long weeks, how could world leaders not versed in the details possibly succeed in less than 24 hours?

How had it reached this crisis? And was there anything Obama—or anyone else—could do to save the day?

Great expectations

The lead-up to the Copenhagen Summit had been hectic. An unprecedented level of diplomatic activity had garnered enormous interest from the media and the public. Many environmentalists were actively raising expectations on what Copenhagen might achieve, hoping to force governments to greater levels of ambition. To them, this was the chance they had been waiting for to take the fight against climate change to a new level. Many news outlets had responded enthusiastically, publishing a steady stream of articles and stories about why a strong outcome in Copenhagen was needed.

The activists and media were right. The scientific case for urgent action was compelling. Less than two years earlier, the Intergovernmental Panel on Climate Change (IPCC)—the world's foremost group of climate scientists—had published its latest major report. Compiled by hundreds of experts and based on more than 6,000 peer-reviewed scientific studies, the IPCC's 2007 report made a powerful case for action. Its main message: global warming's impacts could be 'abrupt or irreversible, depending upon the rate and magnitude of the climate change.' In other words, urgent action was needed to avoid disaster.

This startling scientific consensus was not the only reason for urgency. An earlier treaty, the Kyoto Protocol, had agreed on some initial goals for cutting emissions. However, while it had been warmly welcomed as a breakthrough at the time, Kyoto was now viewed as too limited to fix the problem on its own. For a start, Kyoto's emissions targets applied only to the world's wealthier nations. Furthermore, they were set to expire at the end of 2012. Clearly, a new, more ambitious agreement was needed. Hopes for Copenhagen were running high.

'There was a belief before Copenhagen that we could achieve an ambitious, legally-binding agreement,' recalled Dr Ian Fry when I interviewed him in 2021.[1] Fry, an affable Australian who served as a negotiator for the government of Tuvalu, was an important figure in the climate negotiations, tirelessly representing the needs of small island nations in the Pacific, the Caribbean, and elsewhere. For many of these developing countries, climate change was not just a threat to their economic wellbeing, but to their survival. Sea-level rise could swamp many low-lying islands, spurring many of their governments to call for a binding treaty with the most ambitious goals possible.

As the extent of the danger became ever more apparent in the 2000s, island nations took some eye-catching steps to highlight their perilous position. Mohamed Nasheed, President of the Maldives from 2008 to 2012, decided to hold a meeting of his cabinet *underwater*. Just weeks before the Copenhagen Summit, Nasheed and his fellow government ministers—wearing scuba gear and flippers—assembled beneath the waves around a coral reef, took their waterproof seats and sat down as colourful fish flitted between chair legs, witnessing the bizarre scene. The publicity stunt worked, attracting journalists from around the world to report on what might happen to the low-lying islands of the Maldives if sea-level rise continues unabated.

Such media-savvy political events raised public awareness—and expectations—to an unprecedented degree.

As the Copenhagen Summit approached there were certainly grounds for optimism. The United Nations-sponsored preparations had been thorough. A steady stream of meetings had been taking place throughout 2009, with major gatherings in Bonn, Bangkok, and Barcelona appearing to make some headway in the lead-up to the main event. Meetings of the G7 and other global powers had also been generating positive signals. Furthermore, major governments like those of France and the UK had made clear their commitment to making progress.

Finally, Barack Obama's recent election in the US had boosted hopes that change was possible. Obama's story is so well known it is barely worth repeating: the first African-American president in history, Obama had stormed to victory in November 2008, electrifying progressives around the world with his message of optimism and possibility. In his first few months in office, he had already helped America start to recover from its worst recession in decades, as well as introducing landmark legislation to reform healthcare.

President Obama's stance on environmental issues seemed sensible, especially when compared with his predecessor in the White House, George W. Bush, who had turned his back on the Kyoto treaty and taken what some called a 'do-nothing' approach to climate change. By contrast, Obama had made his belief in mainstream climate science and the need for action clear: 'Just about every scientist ... believes climate change is real ... [and] we can't afford to hesitate much longer,' Obama wrote in his 2006 book, *The Audacity of Hope*. With such an enlightened individual now occupying the most powerful position in the world, environmentalists seemed justified in feeling hopeful.

What could go wrong?

Yet in spite of this optimism, some insiders had misgivings. 'I felt the process wasn't keeping up with the level of ambition,' recalls Richard Kinley, the former deputy head of the UN's climate secretariat.[2] A new international climate agreement would need to cover a wide range of issues, from how countries cut emissions (known as mitigation), to how they might adapt to change once it happens, to how wealthy countries should help out poorer ones. The topics were politically sensitive and incredibly complex. Furthermore, Kinley was concerned some key governments did not seem ready at that stage to take on major new commitments.

'The process had become so complicated in the lead-up to Copenhagen,' says Dr Lisa Schipper, an IPCC author.

> There were so many different groups and sub-groups discussing so many different topics, it seemed like almost no one could keep up with everything. I remember worrying how, with so much on the table, negotiators would figure out where the compromises lay.[3]

Dr Schipper was right. As delegates flew into Copenhagen there were literally hundreds of pages of documents awaiting their approval. Worryingly, most of these pages were littered with 'brackets'—which is UN-speak for disagreements.

Some of these differences appeared daunting. In particular, a growing number of Western governments were eager for developing countries, and especially some of the larger ones like China, to take on more ambitious goals. These rich countries increasingly felt that the 1997 Kyoto treaty, in which only the world's wealthiest nations had agreed to cut emissions, could not deliver the kind of global reductions needed to slow climate change.

The intervening years since Kyoto had only made this problem more apparent. After almost 200 years of European and American dominance as the world's major industrial producers and polluters, in 2005 that tarnished crown had passed to China. No longer could the world's wealthiest Western nations hope to solve the pollution problem alone. Now they would need China, India, and other major emerging economies to play a big part. A future treaty, rich countries argued, would require these new polluters to take on goals of their own.

But in 2009, China and others did not agree. If the Western nations had achieved economic success by burning oil, coal, and gas, they argued, why should others be denied the same path to prosperity? That was like climbing to the top of a bright and shiny tower and pulling the ladder up behind you. If the Americans and other rich countries wanted developing nations to chart a new path to economic growth, then these wealthy few would need to help out the rest by sharing technology and giving much-needed money. Besides, the poorer countries argued, America and Europe still polluted far more *per capita* than the developing countries. Some of these rich nations hadn't even kept the promises they'd made when they'd signed the Kyoto Protocol—pledges that were hardly earth shattering. If the rich world was so worried about global emissions, let them first keep their existing promises, and perhaps take on a few more, before asking the developing countries to shoulder the burden.

To make matters even more complicated, the rich-poor divide was not as simple as it seemed. Neither Western nations nor developing countries were truly united. Some developing countries like China, India, and Brazil, did not want a deal at Copenhagen that might limit their economic growth; meanwhile, poorer nations for whom climate change was an existential threat, like the small island nations, were keen for everyone to do as much as they possibly could to limit greenhouse gas emissions.

For island nations, this was less a question of economic growth and more one of survival. The science said a global temperature rise of anything above 1.5 degrees Celsius could spell disaster. Even 2 degrees could be too much. The developing world—known as the 'Group of 77' countries from Africa, Asia, the Pacific, and Latin America—was deeply divided.

On the opposite side of the negotiating table, wealthier nations were just as fractured. Many European governments felt more could be done to show that the West took its responsibilities seriously. They sympathized with the Group of 77's view that the West had industrialized earliest and carried most of the historic responsibility for climate change. Consequently, the Europeans were more open to taking on stronger Kyoto-style targets before trying to wring concessions from poorer nations. But the European viewpoint was not shared by the Americans, Canadians, and Australians, who were keen to see the Chinese and others come on board with more ambition of their own. Giving China and others a 'free ride' would, they felt, put their own economies at a competitive disadvantage, and would fail to fix the climate problem.

In this respect, the opinion of the world's superpower, America, mattered a lot. The US position was in large part dictated by domestic politics, including the influential fossil fuel industry. Newly elected President Barack Obama wasn't sure he could get the American public—let alone his fellow politicians—onside without China and other major powers shouldering more of the burden. In his enthralling 2020 memoir, *A Promised Land*, Obama recalls his difficulties heading into Copenhagen:

> I couldn't blame environmentalists for setting a high bar. The science demanded it. But I also knew it was pointless to make promises I could not yet keep. I'd need more time and a better economy before I could persuade the American public to support an ambitious climate treaty. I was also going to need to convince China to work with us—and I was probably going to need a bigger majority in the Senate ... and [to] lower expectations.
>
> *(Obama, 2020)*

Obama's concerns in the lead-up to Copenhagen were in fact so strong he admits in his memoir to being on the fence about whether he should even attend. In spite of Obama's private doubts, however, publicly the world's expectations were sky high as the first delegates descended on Copenhagen in early December 2009. Could the Summit deliver on these grand ambitions and deliver a major win for the planet?

Leaks and surprises

The meeting had hardly started when things began to go wrong. While most diplomats were still focused on the lengthy official UN documents that had been developed during the previous months, rumours of shorter, alternative texts immediately began to circulate. Had the Danish hosts drafted a new paper? If so, why did so few people seem to have been consulted?

'We arrived in Copenhagen and were immediately confronted by a leaked document,' recalls Ian Fry. There was a rumour the US had had a hand in the new text, working with the Danish Prime Minister's team. For Fry and others, the leaked text was a problem for two main reasons. First, the text was not 'legally-binding,' meaning it could not be used to hold governments to account later. 'We heard the Danish PM and the US—which didn't want a legally binding outcome in Copenhagen—had drafted an agreement,' Fry revealed during our interview.

A second concern for Fry and many environmentalists was the absence of any reference to limiting global temperature rise to 1.5 degrees Celsius—the limit if many small islands were to survive sea-level rise. Instead, at least one 'alternative' text referred to 2 degrees. This was disastrous because, according to Fry, 'the science said 2 degrees would be the end of Tuvalu.' At 2 degrees, many of Tuvalu's islands would simply sink beneath the waves.

To make matters worse, the Danish hosts seemed to be having difficulties coordinating their position internally. Writing shortly after the meeting, BBC correspondent Richard Black recalls 'tales of a huge rift between [Danish Prime Minister] Mr. Rasmussen's office and the climate department of minister Connie Hedegaard.' What's more, key Danish negotiator Thomas Becker had been removed from his role just weeks before Copenhagen began. Since he had been a key Danish diplomat and a trusted contact point for many other delegations, this threw the talks into initial confusion. Some diplomats were unsure which Danish team—the PM's or his climate minister—was really running things.

Coming on top of rumours of internal Danish divisions, the leaked 'alternative' outcome document caused great uncertainty among arriving delegates. It also spawned still more alternative processes.

'A second group, the so-called Green Group chaired by the Maldives and including Australia, Singapore, Switzerland, Colombia, and others, started meeting in Copenhagen,' recalls Fry. Other groups, including China, Brazil, India, and South Africa, also reportedly began to hold their own parallel processes.

In fact, small gatherings of like-minded delegations were nothing new in the UN system. However, the idea that several possible alternative texts were circulating led to a great deal of confusion. Should diplomats continue to focus on the lengthy official drafts or switch their attention to the shorter alternatives?

'There were parallel processes and mixed signals,' recalls Dr Kati Kulovesi, a former colleague of mine who attended the Copenhagen Summit.[4] 'We arrived in Denmark to huge expectations but no clear view of how governments might deliver on them,' she told me in an interview in 2021. At the time, Kulovesi and I were both working for the *Earth Negotiations Bulletin* (*ENB*). A specialist publication read by diplomats, activists, and academics, the *ENB*'s teams of experts report from negotiations on many environmental treaties. It was—and indeed still is—widely respected for its neutrality and quality analysis of what is really going on behind closed doors at many UN meetings.

What Kulovesi recalls most vividly from Copenhagen was the sense of confusion.

> I became worried when government ministers arrived during the first week of the meeting and began complaining the documents were too complex and full of disputed, indecipherable text. Most experts agree the politicians should be brought in only when there are a small number of outstanding issues left on the table, not when there are hundreds of pages of contested text.

Good intentions… gone astray

The Danish hosts tried to get things back on track by simplifying things. Short of time and eager to secure a deal, they reached out to a handful of key countries with a new text. Their hope was that if some of the major powers could be persuaded to sign off on a streamlined agreement, they would be able to bring the rest of the international community along with them.

The plan backfired spectacularly. They ended up 'annoying every country not on the list, including most of the ones that feel seriously threatened by climate impacts,' the BBC's Richard Black reported at the time.

'The process failed in part because of the small groups that began negotiating behind closed doors,' Richard Kinley agrees. In short, too many governments felt excluded. The UN climate treaty is supposed to be a broad church: more than 190 countries are supposed to sign off on any deal. 'It was a lesson in the need for transparency and inclusion,' Kinley said.

The situation did not improve. As the days ticked by, positions remained entrenched and the tensions mounted. A deal seemed unlikely, if not impossible. As the final day of the Summit dawned, cold and blustery, the outlook was very bleak indeed.

Obama enters the fray

President Obama arrived just when things seemed at their worst. By that final Friday, December 18, a steady stream of Presidents and Prime Ministers from dozens of countries had already touched down in Copenhagen in the hopes of celebrating a breakthrough. By the last morning, however, such hopes had dimmed.

Barack Obama himself had been pessimistic from the start. In his 2020 memoir, he recalls agreeing to come to Copenhagen only after polite but persistent pressure from UN Secretary-General Ban Ki-moon: 'He brought it up at G20 meetings. He raised it at G8 meetings. Finally, at the UN General Assembly plenary in New York in September, I relented, promising the secretary-general I'd do my best to attend so long as the conference appeared likely to produce an agreement we could live with,' Obama recalls. 'I felt like a high schooler who'd been pressured to go to the prom with the nerdy kid who's too nice to reject'.

FIGURE 6.1   President Obama and his entourage arriving at the Bella Center. Photo-graph courtesy IISD/ENB/Leila Mead.

Obama was so unsure Copenhagen could produce anything significant, he postponed his arrival at least once and was virtually the last world leader to arrive at the Bella Conference Centre.

'He didn't look happy to be here,' recalls Kati Kulovesi, who was seated near him in one of the plenary sessions. Still, Obama was determined to salvage some-thing from the chaos. Even at this eleventh hour, he thought he might still be able to secure a softer 'political' declaration rather than a stronger, legally binding one—a compromise he hoped would bring other major powers to the table. Get-ting even a political statement agreed still seemed unlikely, however. According to Obama 'the Europeans were [still] holding out for a fully binding treaty, while China, India, and South Africa appeared content to let the conference crash and burn and blame it on the Americans.'

This was something Obama could not countenance. With Secretary of State Hilary Clinton alongside him, Obama spent a frantic few hours rallying sup-port for his deal. First, he sat down with European leaders—Germany's Angela Merkel, Nicolas Sarkozy of France, the UK's Gordon Brown, and others. After a lengthy back-and-forth, the weary Europeans agreed to Obama's more modest draft outcome … provided he could get the 'big four' developing countries—the Chinese, Indians, Brazilians, and South Africans—on board.

A party-crashing president

But where were they? Obama's staff couldn't find them. Were they avoiding him? Eventually, Obama and Clinton discovered them upstairs holding a meeting of their own. In a small room behind closed doors and tight security were Chinese Premier Wen Jiabao, President Lula of Brazil, Prime Minister Singh of India, and South African President Zuma.

In Obama's version of events, he 'crashed… [their] party' and sat down uninvited with the four leaders, negotiating on the fly as he tried to secure an agreement. He eventually cajoled them into going along with his document, warning them that if they refused and the meeting ended in failure, he would go straight to the international press with his side of the story, ensuring they got the blame. After half-an-hour's haggling and a few final changes to the text, an outcome was agreed.

Obama took the document, strode back downstairs, summoned the attending media, and announced a deal. He left the conference feeling 'pretty good' at what he'd achieved. He had secured what he felt was a 'stepping stone' to future action both within the US and internationally, while persuading China and India to accept that 'every country, and not just those in the West, had a responsibility to do its part to slow climate change.' He left the Bella Conference Centre moments later, his car whisking him back to the airport where he boarded his plane and took off, narrowly avoiding an incoming snowstorm and arriving back in the US a few hours later.

Job done, right?

Actually, no. President Obama may have persuaded China, India, the Europeans, and some others to grudgingly go along with his deal. The trouble for him was that more than 150 other countries were part of the climate negotiations. For it to be a 'real' internationally agreed outcome, they would all have to sign off on it, too.

A nightmare on Bella Street

Participants in the Bella Centre were in a state of shock and surprise. Had they just seen Obama hold a press conference and claim a deal had been made? Weren't they supposed to be asked, too?

Danish Prime Minister Rasmussen evidently considered the deal was already done. After all, it sounded like the leaders of the world's other major powers, from the Chinese and the Indians to his own European colleagues, had all agreed to it, however grudgingly in some cases. It might not be as strong on commitments as the European Union wished, and it might mean developing countries would need to take on their own commitments down the road. But still, a deal was a deal. The Copenhagen Accord, as it was called, had been agreed.

Or had it?

When Rasmussen opened the final plenary session and tried to gavel the agreement through, a shock was in store. While many exhausted diplomats signalled their acceptance of the Copenhagen Accord, a handful were vehemently opposed. Bolivia, Venezuela, Sudan, and Tuvalu all registered their opposition in the most passionate language. Stunning her fellow delegates, Venezuelan diplomat Claudia Salerno Caldera held up a hand dripping blood while declaring her country's opposition to Obama's deal. Caldera had, it turned out, been banging her hand against the table so hard in an effort to attract the Chairman's attention that a ring she was wearing had cut deeply into her hand, causing blood to flow—something she had only realized when she was speaking to the assembled group. And in spite of pressure from more powerful nations, Tuvalu's Ian Fry gave an impassioned speech for a stronger treaty, bringing cheers from activists and environmentalists at the back of the packed hall.

Although most government delegates now seemed willing to accept Obama's document—some happily, others grudgingly—there was too much vocal opposition in the room for Rasmussen to ignore. Under the rules governing the UN's climate meetings, consensus was needed. With several delegations still opposing the deal, they had hit an impasse.

Talks continued throughout Friday night and into Saturday as exhausted delegates tried to salvage something. Richard Kinley recollects increasingly irate diplomats, many usually so suave and calm, finally losing their tempers. One became so incensed he began screaming at the UN Secretary-General to 'do something!' to save the meeting. 'The last few hours were the worst experience of my career,' Kinley recalls.

In fact, Kinley was one of the few who managed to keep his composure. 'He was calm and normal when the pressure was getting to everyone else,' Kati Kulovesi remembers. In the midst of the chaos, Kinley and other key UN staff somehow stayed unruffled as they explained the options available to stressed and sleep-deprived ambassadors and politicians.

Tempers were boiling over when UK Climate Minister Ed Miliband took the floor and suggested the meeting be adjourned for a short time. During this break in proceedings, participants had time to draw breath and calm themselves. Behind the scenes, UN Secretary-General Ban Ki-moon reportedly met with several delegates, including those who seemed the most furious about what was happening. As unflappable as Kinley, Ban Ki-moon's efforts to defuse the situation apparently helped. Meanwhile, the exhausted Rasmussen allowed one of the conference vice presidents to take the chair, allowing for a fresh face on the podium. When the meeting resumed early on Saturday morning, much of the anger had dissipated.

With delegates seemingly in less fractious mood, thoughts turned to finding a way to break the impasse. Even with the mood of the room improving, however, there remained a serious difference of opinion on whether Obama's proposed document should be adopted by all governments present.

It was around this time that a possible solution emerged. If governments could not all endorse the agreement, could they not at least 'take note' of the outcome? This would still give the document some level of legitimacy and keep it on the table for future meetings, even if it did not signal full agreement.

The suggestion did the trick. On Saturday afternoon, a full day after its originally scheduled end, delegates agreed to 'note' Obama's document and the meeting wrapped up. Bleary-eyed delegates began to drift away, zombie-like, from the plenary.

But what had they achieved, exactly? Had the Summit ended on a high or was Copenhagen a dismal disaster?

A successful failure?

At the time, Copenhagen definitely felt like an unmitigated failure. I was there. Even today, my memories of the meeting are unrelentingly negative. Closing my eyes now, I vividly recall navigating the corridors, many filled to overflowing as 30,000 people try to fit into a space designed for 15,000. I remember long lines of delegates waiting outside in the cold and snow to get through security, interminable meetings, and more than one diplomat losing their usual composure and engaging in shouting and abuse. I see my talented *ENB* colleagues voicing their fears and disappointment, my friends among the UN staff, diplomats, and activists venting their frustration. I remember working past midnight night-after-night as we wrote and published our reports, fuelled by increasing amounts of caffeine and running on adrenalin. I recall my sense of disappointment and even depression in the days and weeks afterwards.

In an email to a close family member just 24 hours after the meeting ended, I wrote: 'Copenhagen was exhausting and … upsetting. People here are either describing it as a "first step" or as "weak." But I've already heard the phrase "first step" describe Rio, Kyoto, and Bali before. How many first steps do we need?' In my private diary entry written a week later, I write of 'a weak outcome' and 'angry delegates,' describing it as 'a meeting that failed to live up to its billing and only narrowly avoided total collapse.'

Friends and colleagues from the time tell me I was far from alone; many felt a sense of sadness for the process, of a missed opportunity for the planet. Richard Kinley, for instance, remembers his colleagues leaving Copenhagen 'devastated and demoralized.' Dr Lisa Schipper recalls being 'not surprised … but still shocked and disappointed.'

In recent years, though, the view of Copenhagen has changed. While many observers would still consider it an organizational and procedural failure, most now believe it achieved significant substantive success.

Even on the organizational side, it provided key lessons on how to run—and how *not* to run—the next round of climate talks. The BBC's Richard Black recalled Copenhagen as a procedural 'farce'—something that should not be repeated. Copenhagen certainly taught host countries the critical importance of transparency, Richard Kinley argues. 'After Copenhagen, it became impossible to have small [closed] group negotiations,' he says. For a truly multilateral treaty to make progress, all countries had to feel engaged and informed. Having learned some hard lessons in Copenhagen, the hosts of the watershed Paris Summit six years later became a 'poster child of organization,' Kinley believes.

Second, Copenhagen is now widely considered to have delivered some genuinely substantive outcomes. For a start, industrialized developed countries pledged to provide developing countries with US$100 billion in financing annually by 2020 to combat climate change—a significant promise on which wealthy nations can (and should) be held accountable. A 'Green Climate Fund' was formed shortly afterwards to support this undertaking.

The Copenhagen Accord also included an acknowledgement from developing countries that they, too, would need to take on climate mitigation goals and report back on their progress to the UN. This recognized that the problem had become too large for the West to deal with by itself. Ultimately, this idea of a shared responsibility evolved into the idea of national targets—or Nationally Determined Contributions—adopted in Paris in 2015.

Finally, the Copenhagen Accord highlighted the need to limit global temperature rise to below 2 degrees Celsius. While this was considered too high by many activists, a softer reference to 1.5 degrees was also included in the text, at the urging of many developing countries. All three of these substantive outcomes, which Copenhagen merely 'takes note' of, would emerge much more powerfully as part of the Paris outcome.

Did Copenhagen help pave the way to Paris? Should a Paris-type outcome have happened a lot sooner? Quite possibly. Looking back, though, it is quite likely Obama and others were right: the world may not have been ready in 2009 to take such a major leap forward. China, India, and other developing countries clearly needed time to digest the reality that all countries would need to take on commitments. President Obama *admits* in his memoir that he wanted more time to win over crucial domestic allies. By the time Paris came around six years later, though, it would be a different story. Those early seeds from Copenhagen, carefully nurtured in the intervening years, would finally grow and bear fruit in Paris; something the next chapter will show.

Will the real hero please step forward?

Looking back with the beauty of hindsight can we really say there was a 'hero' at the Copenhagen Summit? If so, who was it?

Some would argue for Obama. After all, he managed to persuade other world leaders to agree to his deal, which included some new and important ideas that would finally blossom in Paris. Obama's last-minute diplomatic efforts, his energy and his sheer determination to find a solution deserve praise. How many other presidents would have been willing to engage in a high-level game of hide-and-seek and been willing to barge in on other world leaders holding a private meeting, then sit down with them and hammer out a deal? Without Obama, the Copenhagen Accord would almost certainly not have happened.

But Obama does not seem to be a universal choice. In fact, some might even characterize him as the villain of the piece. 'The immediate reason for the failure of the talks can be summarised in two words: Barack Obama,' wrote *Guardian* columnist George Monbiot just days after the Summit. 'Obama cannot be proud of what he achieved in Copenhagen,' Ian Fry opines. 'We could have achieved more there.'

If Obama is not everyone's choice of Copenhagen hero, who else might be?

What about Richard Kinley and his talented colleagues in the UN? Kinley in particular remained unflappable and unruffled while many others seemed to be losing their heads. 'Perhaps he was the hero,' Dr Kulovesi suggests, 'for remaining calm and making sense when everything had descended into chaos.' Kinley's composure in a crisis, his patience in explaining UN rules and procedures to exhausted delegates on the final day and night, undoubtedly helped avoid a total breakdown in the talks and, perhaps, ensured they finally limped over the line.

Or how about Ian Fry? In the face of intense political pressure from powerful governments, and with the support and affection of environmentalists everywhere, he and a handful of others stood up for environmental integrity and refused to back away from his belief that 1.5 degrees should be the acceptable limit of global temperature change. Could his principled stand earn him the label 'hero of Copenhagen'?

The reader will need to make their own mind up. What we know for sure is Copenhagen did not deliver the breakthrough many had hoped. For the next chance to make genuine progress, climate watchers would have to wait another six years, until a new day dawned in Paris.

Notes

1 Interview with Ian Fry via video conference, April 28, 2021.
2 Interview with Richard Kinley via video conference, April 1, 2021.
3 Interview with Lisa Schipper via video conference, May 14, 2021.
4 Interview with Kati Kulovesi, April 15, 2021.

Background References

Akanle, T., Appleton, A., Kulovesi, K., Schulz, A., Sommerville, M., Spence, C., and Yamineva, Y. (2009) Summary and analysis of the Copenhagen climate change conference, *Earth Negotiations Bulletin*. Available online at: https://enb.iisd.org

Black, R. (2009) Why did Copenhagen fail to deliver a climate deal, published on *BBC News* website, Available online at: http://news.bbc.co.uk/2/hi/8426835.stm

Depledge, J. (2010). The outcome from Copenhagen: At the limits of global diplomacy. *Environmental Policy and Law*, Vol. 40, no.1, pp. 17–22.

Monbiot, G. (2009) If you want to know who's to blame for Copenhagen, look to the US Senate, published in *The Guardian* online, December 21st 2009. Available online at: https://www.theguardian.com/commentisfree/2009/dec/21/copenhagen-failure-us-senate-vested-interests

Reference

Obama, B. (2020) *A Promised Land*, Viking Press.

7

CHRISTIANA FIGUERES

The Can-Do Advocate: The Paris Agreement on Climate Change

Andrew Higham

This is the story of a hero who made heroes of us all and repurposed the greatest threat to humanity into the greatest opportunity for the flourishing of life on Earth. A movement we call the Great Regeneration (Mission 2020, 2020).

In the European Spring of 2010, in the aftermath of what was then regarded as a shocking failure of the Copenhagen Climate Summit, the United Nations Secretary General appointed Christiana Figueres as the new Executive Secretary of the United Nations Framework Convention on Climate Change (UNFCCC).

Karen Christiana Figueres Olsen was born in San Jose, Costa Rica in 1956. Her father was the great Costa Rican revolutionary politician Jose Maria Figueres Ferrer who led the nation to become a shining star of modern democracy. Three times President during the period from 1948 to 1972, he transformed Costa Rica's education and health systems, revamped its economy, abolished its standing army and oversaw the creation of an outstanding system of protected areas throughout the country in recognition of its natural and cultural heritage. Figueres's mother, Karen Olsen-Beck, was also a formidable champion of Costa Rica, having acted as Ambassador to Israel and an elected representative of the Legislative Assembly. Figueres's childhood was therefore steeped in the politics, demands, and privileges of a revolutionary family and Head of State, which was a crucible that forged her as an intensely determined force of nature. And with that foundation, she embarked upon her own career in diplomacy, yet with a uniquely diversified experience ranging from anthropology, to lobbying, to organisational leadership and change, and to the power of the non-governmental and private sectors as drivers for social change and climate action.

Figueres joined the UNFCCC Secretariat as a stalwart of the UNFCCC climate regime. She had been the leading negotiator for Costa Rica and the Latin American region for over 15 years, a UNFCCC Bureau member and Vice

DOI: 10.4324/9781003202745-8

President, had represented her region on the Clean Development Mechanism (CDM) Executive Board, and as such she was well acquainted with all aspects of the climate change *problematique* and the key actors in the field. She had a deep understanding of why Copenhagen had collapsed (as well as where progress had been made and what could be harvested in the years ahead) and what it would take to rebuild the climate regime. In an interview upon the announcement of her appointment she said, "I think there's a sense of hopelessness, that we're not getting anywhere, and we really need to turn that around. We really need to create an ambiance of 'Yes, we can do that.'"[1]

Even prior to her official entry into duty, she visited the demoralised seat of the UN climate Secretariat in Bonn for a workshop in the garden of Haus Carstanjen, the historic building that housed our staff. In an effort to motivate her new team, at her insistence we literally all painted visionary abstract artistic impressions of what success would look like when we had saved the world from the climate crisis. We were invited to dream big. And then we went back to work. She knew that she had to rebuild from the core out into concentric circles. As she has stated, her first and second priority was to reinvigorate the Secretariat and her third was to empower collective leadership and to do so by building ever-increasing circles of allies[2]. We would have to convince ourselves first before we could convince the rest of the world. Her early visit to Bonn to reinvigorate the Secretariat was just the beginning of her campaign to reinvigorate the entire regime.

Impossible is not a fact. It is an attitude

Soon after Figueres moved into her office overlooking the garden where we had painted our visions of success, she marked it with a motto for herself and anyone who would come to meet her. As the sign said, "Impossible is not a fact. It is an attitude." At every opportunity she spread the "can do attitude." As later explained in The Future We Choose (Figueres and Rivett-Carnac, 2020) she knew her primary task was to "be a beacon of possibility that would allow everyone to find a way to a solution together." Figueres was relentless about it and would not tolerate the faintest fatalism within the UN system and beyond. She knew that if the negative self-talk of the climate community was to prevail, there would be no winning the climate crisis. So she proactively countered the proliferation of negativity among "friends and family" – networks of climate actors from civil society and other stakeholders, private sector, UN constituencies, and governments – whether on the phone, in person or through the media.

Through the multilateral process, which the UN Climate Secretariat's deputy, Canadian Richard Kinley, had deftly led for so long with superb strategic substantive guidance from another senior staffer from Iceland, Halldor Thorgeirsson, the 16th meeting of the Conference of the Parties (COP) in Cancun in 2010 was the essential moment to harvest the lost gains from Copenhagen in 2009 and rebuild trust in the multilateral process. It was critical that the COP build trust from developing countries in particular, by mobilising the capacity building, financial and technology development, and transfer support that would become fundamental to progressing towards a legally binding agreement applicable to all Parties. Only movement on these issues would persuade the Group of 77 (UN parlance for developing countries) to accept stronger commitments as a part of any future agreement.

The rebuilt foundations of Cancun enabled the South African government, which was hosting the next meeting in 2011, to reinitiate a focus on this "bigger picture,"[3] as they began to describe it, while also delivering on the pre-2020 matters of priority (full implementation of the Bali Roadmap[4]). Through a series of Ministerial dialogues starting in Bonn and ending at the pre-COP in Stellenbosch,[5] the South African Presidency created a clear pathway for the conclusion of the past era of climate diplomacy and the beginning of a new era as expressed through the Durban Mandate, which made clear the path (and ambition) to adopt a new legal agreement in 2015 which would come into effect from 2020. South African lead negotiator Sandia de Wet, Indian legal expert Lavanya Rajamani, and I drafted the Durban Mandate through the Indaba process[6] while the chaos and drama of COP17 spiralled all around us. While the Durban Mandate became the primary decision outcome of COP17, the attention of most negotiators had not yet shifted towards planning for the successful adoption of a new legally binding agreement. For the most vulnerable nations, waiting until 2020 for a new agreement was untenable. The deal in Durban therefore also saw the creation of a new and much more creative and practical "workplan on enhancing mitigation ambition" (UNFCCC, 2011) focused on implementation and bridging the gap in ambition by 2020 through a range of specific measures and international cooperation to reduce emissions. This was the seed that has since transformed the UNFCCC into a space for climate action by non-state actors, recognised at the heart of the Paris Agreement's DNA, and now the overwhelming focus of the multilateral climate process.[7]

Bring out the best in everyone

Figueres loved bringing her knowledge and passion for organisational change to effect in her new role. Within weeks of taking the job, she had met every single member of her 500+ staff, providing everyone with her mobile phone number and insisting on an open-door policy for the entire team. She fostered a deep organisation-wide camaraderie, and she made sure that she was personally engaged in the lives of her team and instilled a culture that expected all to bring the best out of each other. She brought a fresh equality to staff relations, with all being treated with the same respect and generosity of spirit – whether administrative staff, local staff and cleaners, directors, or professionals. She opened her home to staff members, she encouraged regular social activities and, herself a long-distance runner, she actively fostered the importance of keeping fit and healthy. She knew this was going to be essential if we were to achieve our vision –

it was after all a marathon of sorts. The workload of the UNFCCC Secretariat was intense, even outside of negotiating sessions. During negotiating sessions staff would rarely get more than a few hours' sleep each night over the course of weeks. Figueres organised food hampers for staff stuck late in their office at night, she turned offices in negotiating sessions into massage clinics, to express her gratitude she sent personal messages to the individuals who worked incredibly hard yet often received little acknowledgement. She brought the best out of everyone and created a reciprocity of kindness that reverberated so that *everyone* started to bring out the best in everyone else.

Genuine infectious enthusiasm and love for life

After a period as Secretary of the Ad Hoc Working Group on the Durban Platform for Enhanced Action (the negotiating forum governments created as part of the Durban Mandate to negotiate the Paris Agreement), I was asked to take on broader responsibilities for the Paris Outcome in the role of Senior Advisor and Manager of the Drafting and Advisory Team. This team was responsible for supporting all negotiations across all the elements of the emerging Paris package. In an organisation that had a habit of compartmentalising negotiating support by agenda item or negotiating body, it was a rare opportunity to build a single Secretariat-wide team that was dedicated to delivering a fully integrated overall Paris outcome, with the Paris Agreement treaty text at its core. Without Figueres's determination that would have been impossible. Because we built such a strong and dedicated team that was driven to deliver the Paris Outcome at the highest possible ambition and without a whisper of doubt, we worked with a great intensity. Yet Figueres also reminded us of the importance of balance, and with an infectious enthusiasm and a sheer love of life, she kept us grounded. As she did one shiny Bonn Autumn day in 2015 – having ridden her bike past a glorious hedge of crimson ivy vine along the Rhine, once she arrived at work she raced along the corridor where my team worked, enamoured by the day and insisted we drop everything, get on our bikes, and take the morning off and witness the beauty all around us.

Planning for success in Paris

I may not ever again meet such an extremely intuitive strategist as Figueres, especially one with that degree of emotional intelligence and self-awareness. She used a visionary leadership style and gave agency to develop and implement strategy: for the negotiations, the broader political dynamics, communications, and the mobilisation of influencers and non-state actors. Her activist and interventionist approach was highly respectful of Parties to the Convention[8] and prudent given the setting of a United Nations treaty body. Yet she also intentionally pushed the boundaries. She would remind us that the 'atmosphere was our ultimate guide' (referencing the objective of the Convention). It was a rare opportunity as a servant of the Parties to a Convention to be mandated to actively shape political solutions and develop ideas that would manifest as textual landing grounds for a treaty outcome that was to be universally adopted. By 2012 we had a detailed yet constantly evolving map of the negotiation processes that would result in an ambitious adopted outcome in Paris. By 2014 the Drafting and Advisory Team had developed a highly sophisticated textual map of outcomes across the vast terrain under negotiation. Figueres also recognised that the UNFCCC Secretariat could be stronger in political strategy and while she engaged those across the organisation with the best political brains, she also engaged trusted allies beyond the organisation and bolstered the organisation by creating in 2014 a separate small philanthropically funded and covert Groundswell Initiative, headed by Tom Rivett-Carnac.

Groundswell was a campaign that aimed to support our efforts to build an ambitious Paris Outcome and create a positive 'mood music' that would ensure that national governments perceived the adoption of the Paris Agreement as inevitable and highly desirable. They generated 'surround sound' by mobilising a broad array of non-traditional voices for climate action. Along with the UN Secretary General's Climate Action Support Team and the Lima-Paris Action Agenda, they orchestrated a crescendo of announcements by businesses, investors, and subnational actors that gave a sense that momentum for action on climate change was building and that governments would be supported by these actors in taking on ambitious commitments in Paris. As a part of the Groundswell Initiative, Figueres established Strategic Councils that were vehicles to engage high level individuals from many different constituencies and countries, including the leadership of faith organisations, investors, businesses, subnational governments, military organisations, and many more. She also regularly engaged a wide range of existing networks. Increasingly throughout the year leading up to Paris she used these Councils and networks to make explicit asks of leaders to intervene directly, to make public statements, to make private phone calls, or to help deliver a specific outcome that was important for a country to become more enthusiastic about an ambitious outcome in Paris. The Groundswell Initiative also coordinated with the increasingly well-coordinated and effective civil society campaigning lead by networks of non-governmental organisations such as Climate Action Network or the Climate-Sustainable Development Goal Coordination Group. Of note was the Climate Briefing Service and the role of Jennifer Morgan, who applied incredible leadership and wisdom in guiding civil society through her "über-strategist" role.

Figueres' success was in making the success of Paris everyone's job and in everyone's interest – she created the beacon and shared the light with everyone who we needed for Paris to be a success. "All hands on deck," as she would say. Everyone was welcomed to the party, everyone assumed or was given a task, or encouraged to contribute to a successful outcome, even the CEOs of some of the world largest energy companies and historic polluters. Many who were central to these coordination efforts across the various nodes of the broad climate movement reflect on this time leading up to Paris by high-lighting how beautifully the movement functioned, like an ecosystem or some super-high-performance team.

I recall one mid-2015 meeting of the innocuously titled 'Implementation Coordination Committee', which was the official senior management forum through which the UNFCCC Secretariat coordinated its substantive strategy for the negotiations. When assessing our climate action strategy, it was noted that climate action initiatives were proliferating exponentially and at such an out-of-control scale and pace, it was no longer possible to keep track of them. That was taken as an ultimate measure of success.

The campaign mindset

Figueres brought a campaign mindset to the role of Executive Secretary. While it was welcomed by many in the field, it did not sit entirely comfortably with some of her more traditionalist UNFCCC Secretariat colleagues. She brought that mindset thanks to her political pedigree and learnings through her life, but perhaps quite specifically from her experience as a lobbyist leading the international campaigns for the Hawthorne Group, among other roles.

Within the UNFCCC Secretariat her "one team" approach required the internal silos to be broken down, threatening small power and political comfort zones and tribes. Similarly, her call to a higher purpose and complete determination ("impossible?...wrong attitude!"), was not always greeted with enthusiasm, and her pushing the boundaries of what was appropriate within a UN bureaucracy required an inoculation to frustration. She had mobilised plenty of good will within the Secretariat to win those battles, yet within the UN System the campaign mindset was most valuable since proxies for the negotiations abounded. For example, the notion that the Paris Agreement would be enduring and enshrine a long-term goal was challenged within the UN System. We had to fight hard for this essential pillar of the Agreement's ambition to be embraced, in particular, the "net-zero" language now in Article 4 and the triptych goal (related to mitigation, adaptation, and finance) now set out in Article 2 of the Paris Agreement. Ban Ki-moon would become a strong advocate for the agreement and indeed his Summit held in the UN Headquarters in New York in September 2014 made a valuable contribution in socialising key elements of what would become the architecture of the Paris Agreement, which was captured in a Chair's Summary issued upon conclusion of the Summit.[9]

While an intuitive strategist who empowered as a "beacon" and avoided micro-managing, Figueres did instil an attitude of planning for success. We left no scenario to chance, having gamed the negotiations thoroughly, covered all bases in our campaign strategy, and thought through and designed text for all possible landing zones. We applied the scout's motto of being prepared in earnest, much to the discomfort of the Drafting and Advisory Team members. And we did so in the most covert manner possible, as the Secretariat is most effective when it is invisible and the negotiating process flows naturally, with Parties always feeling directly in control. In servicing the Parties, the Secretariat does so through national representatives elected as presiding officers, who share with the Secretariat the custodianship of the negotiating process. In October 2015, that custodianship broke down, when the Ad Hoc Working Group on the Durban Platform Co-Chair's rejected the advice of the UNFCCC Secretariat and the French and Peruvian Presidencies and issued a text for negotiation that was far from where Parties had matured their negotiation (IISD, 2015). The Co-Chairs so badly read the mood of Parties that they may have lost the legitimacy of their Chairmanship in that session. After negotiators regained ownership of the text, they began to explicitly seek for textual proposals from the UNFCCC Secretariat, rather than from the Co-Chairs, and the last session before Paris concluded with a mandate for the Secretariat and not the Co-Chairs to prepare an input for negotiations in Paris. Yet this was not perceived as a real setback. Nothing was going to impede the incredible momentum that had been generated through the campaign. Indeed, Parties had already triggered, through the Geneva session in February 2015, the formal requirement for a treaty outcome in Paris and capitals were all preparing for this outcome. The real question was how ambitious could it become? Fortunately, the French Presidency, led by Laurence Tubiana and her right- and left-hand masterminds, Emmanuel Guerin and Anne-Sofie Cerisola, had laid careful groundwork for that through their brilliant diplomacy and well-curated consultative processes.

The beautiful text

During the year leading up to Paris, Richard Kinley convened white board strategy sessions among the core Secretariat team responsible for success in Paris, where the political design of the Summit and management of COP preparations occurred. After the Co-Chairs fatally self-harmed their chairmanship with their October text, the need to advance textual alignment with the French Presidency became a most important priority, and I was dispatched to Paris with my colleagues Daniel Klein and Conor Barry. One could not be too careful given the sensitivities associated with Presidential texts at this critical point in the negotiations. Most painfully, the spectre of the Danish text that was leaked days before Copenhagen loomed. It was always clear that there was no French text. However, there was a Secretariat text that had to be protected and that consultation with France proved to be of great benefit in aligning our understanding of what was achievable and where the true red lines of Parties were, in preparation for the inevitable handover between the Ad Hoc Working Group on the Durban Platform for Enhanced Action (ADP) Co-Chairs and the French Presidency that occurred on 5 December 2015.

While the consultations continued in various formats and across all the elements of the Paris Outcome into the second week of COP21 in Paris, the French Presidency and the Secretariat worked line by line on the screen, while reports on progress and sticking points in the negotiations came through. Around the table sat the core drafting team, the COP President, Figueres, Laurence, Emmanuel, and Anne-Sophie, along with the Peruvian Presidency (Romulo Acurio and Rosa Morales) who at the invitation of France had remained closely connected to the consultation process throughout 2015. That 24-hour operation lasted from Tuesday, 8 December until Saturday, 12 December, when that evening the Agreement was unanimously adopted by acclamation (and tears of joy). Full draft versions of the Paris Agreement were produced as proposals from the President on 9 December and on 10 December.

Throughout that period the French Presidency engaged in bilateral meetings at which they asked Ministers to identify solutions to the "crunch" issues that remained, and to specify the one or two absolute "must-have" outcomes in the text. They were not asked what they could not live with, as this would have dragged down the overall ambition of the outcome. Through this engagement, an upward spiral of ambition was created in the text, enabling the Presidency to assemble a package that contained the fine balance of ambitious elements that all Parties needed and to gradually improved the outcome through the iterations of 9, 10, and 12 December. Throughout the process, the Secretariat worked hand in glove with the Presidency.

Behind the scenes, the 60 members of the Drafting and Advisory Team, already sleep deprived from ten days of constant negotiations, made this happen like some wonderous organic machine working in shifts 24 hours a day. That entailed managing an array of negotiating settings across the 29 Articles of the treaty and many other specific issues under negotiation, finding solutions to complex textual problems that would satisfy 197 Parties, and producing text on time to a near degree of technical perfection. The document management team was led by Stelios Pesmajogolou. As champagne flowed through the Conference rooms after the Paris Agreement was adopted, these professionals continued working until the next day to ensure that the agreements reached were correctly documented.

As Figueres has noted these document management professionals made Paris a success in equal portions as herself, or the Heads of State that attended the COP, or the activists who demanded that nations raise the ambition of their commitments in line with science, or the UN Security Guards that managed the safety of delegates (not forgetting that Paris had just days prior to the opening of the COP been the subject of a devastating terrorist attack).

In Durban a few years earlier, a "Durban Alliance" had formed between the world's poorest countries (known as "Least Developed Countries"), small island nations, the EU, and progressive Latin and African governments. That group was determined to drive as ambitious an outcome as possible, aimed at holding global temperature increases below 1.5 degrees Celsius. In the lead-up to Paris this alliance reinvented itself as the High Ambition Coalition (HAC), this time led by the Republic of the Marshall Islands' formidable late Minister Tony de Brum and his team, particularly Dean Bialek. The HAC's objective was to rally support from Parties. By the second week of Paris, it was coordinating over 30 Ministers and over 100 nations to secure key elements of the deal, including the 1.5 degrees Celsius temperature goal, the net zero global emissions pathway by the second half of the century, and a five-year cycle for updating mitigation contributions.

The successful inclusion of all of these elements of the deal in the text that was adopted on 12 December was a genuine surprise for many negotiators and observers, in particularly by making the objective of the Agreement to pursue 1.5 degrees Celsius as an addition to the reference to holding global temperatures to "well below 2 degrees Celsius." This was made even more powerful with the inclusion of a request to the Intergovernmental Panel on Climate Change (IPCC) – a group of the world's top climate scientists – to prepare a special report on the impacts of global warming of 1.5 degrees Celsius above pre-industrial levels and related global greenhouse gas emission pathways, in the context of strengthening the global response to the threat of climate change, sustainable development, and efforts to eradicate poverty. The IPCC report elucidated what was at stake between a world at 1.5 degrees Celsius and one at 2 degrees Celsius and enabled the moral compass to be firmly reset at the 1.5 degrees Celsius level of ambition. Ultimately, because of this it became impossible for non-state actors to anchor their commitments to the 2 degrees Celsius goal and within six months, all commitment frameworks of business, investors, subnational governments shifted to this higher level of ambition. In tandem, governments began to do the same in their long term strategies for reaching net zero and in their upgraded commitments (known as nationally determined contributions or NDCs). In late 2021, the UK hosts of COP26 made "keeping 1.5C alive" as the strap line for the entire build-up to the COP. To thank for that we have hundreds of heroes from the most vulnerable nations, the scientific community, and campaign groups who have struggled tirelessly for decades to make 1.5 degrees Celsius the acceptable upper limit of global warming since that is the absolute maximum required if the most vulnerable nations are to survive the climate crisis.

Thousands more share in the success of Paris. In recognition of this inalienable fact, for the anniversary of the adoption of the Paris Agreement in 2016, Figueres initiated with Richard Kinley, and with the help of Tom Carnac and me, the publication of *Profiles on Paris*, which celebrated diverse contributions towards the Paris Agreement. As it states:

> The Paris Agreement of 2015 is a remarkable example of collective leadership. No one individual, institution or country can claim to have singlehandedly created the Paris Agreement. Rather, over several years, thousands of individuals and organizations came together to contribute to the moment where 195 countries unanimously adopted the text.[10]

The even richer history of how the Paris Agreement was negotiated, and the characters that made it possible, was also documented in a book that was published by Oxford University Press in 2017 (Klein et al., 2017).

This is just the beginning

Early in the morning on 12 December 2015, the French and Peruvian Presidencies and the core Secretariat team sat in a circle around my desk in Le Bourget conference centre with printed copies of the final text to do a final read through to ensure that there were no mistakes. We had already pulped 5,000 copies of the text because one block of text transgressed China's red lines, and unknown to us at this moment, the text contained a mistake of having included a 'shall' where a 'should' belonged in Article 4.4[11] of the Agreement – something that the United States could not live with as was well known to us all as it would have required new domestic laws to be agreed in Congress before the Treaty could be given effect.

The text was made available to delegates at around 1.30 pm on 12 December with that error on board, while I slept for a few hours, only to be woken with the drama of this situation beginning to unfold. I walked the plenary floor apologising to delegates and explaining that the text contained a typographical error in Article 4.4, and that a correction would be made along with other minor typographical errors upon its introduction when the plenary got underway. Richard Kinley, the mastermind of the management of the closing plenary and the unfolding situation maintained calm and made the necessary arrangements. Figueres and the French Presidency in the meantime managed the political fallout from the situation shuttling between Ministers. Chinese Minister Xie Zhenhua was particularly instrumental in preventing the situation from spiralling out of hand within the Group of 77 and China. And with a final set of risks from the Like Minded Developing Countries[12] addressed, we took our places on the podium.

The moment of adoption was euphoric and intensely emotional for all those assembled in the plenary room, for those in spillover rooms in Le Bourget and for those watching on live stream around the world. That is the type of joy that lasts a lifetime. I remember looking out and seeing the delight and deep satisfaction in the faces of colleagues, of embracing Figueres, Richard, Halldor, and Lawrence, and the sensation of closure after the long grinding road from the collective failure of Copenhagen to the invigorating moment of collective achievement in Paris. After all the speeches and rounds of embraces, the bottles of champagne flowed, and many danced until morning. And after some sleep came the call from Figueres on Sunday morning, still bubbling with excitement yet also reflexive as her mind turned to the challenges of Monday morning. *We did it! AND now we have an even bigger challenge before us. We must continue to build on the momentum of last night if we are to deliver on this great achievement.*

That marked the creation of Mission 2020, a new collective campaign to channel the energy that had created the Paris moment into the next phase of global climate action. The world now had a mechanism to bend the curve in emissions from 2020 onwards. But the success of the Paris Agreement would depend on whether we could make up for lost time before 2020, reach the climate turning point, and prove that that "ratchet mechanism" of the Paris Agreement worked as intended: all nations and all investors, businesses, subnational governments, and citizens stepping up with more ambition by COP26 in Glasgow to get us as close to a below 1.5 degrees Celsius trajectory as possible.

In the lead up to Copenhagen, the World Bank issued a report that examined the implications of a world that had warmed by 4 degrees Celsius, making the case for why emission reductions to bring the world closer to 2 degrees Celsius were needed. By Cancun, some additional progress was made, and by Paris, the world had advanced to an optimistic scenario of 2.9–3.4 degrees Celsius this century if all emission reduction commitments were fully implemented. In the lead-up to COP26 in Glasgow, the new Biden Administration convened a Summit in May 2021 and encouraged all nations to come forward with greater emission cuts. As a result, the optimistic scenario became 2.4 degrees Celsius this century. And then after the first week in Glasgow, that optimistic scenario has been further improved to 1.8 degrees Celsius this century, albeit noting that based on 2030 targets alone (i.e. excluding the aspirational long term plans of governments to reach net zero by 2050) the temperature scenario was still stubbornly fixed at 2.4 degrees Celsius this century. Because of this gap in ambition, Glasgow agreed that all countries should once again come back to the table in late 2022 with a further revisiting of targets to bring this 2030 projection much closer to the 1.5 degrees Celsius scenario.

The genuine and also strategic magnanimity and humility of Christiana Figueres, Hero of the Paris Agreement, made heroes of all who have made solving the climate crisis our predominant concern, and who see the solutions to climate change as generating a bounty of developmental progress and as the beginning of a new enlightened era for humanity. Figueres has emphasised that climate action must not be an end but rather a means towards achieving better human development, and moreover that climate action was the *only way* that we could preserve and expand the development gains that humanity has achieved (Figueres, 2020).[13]

Today, Figueres continues her relentless and stubborn climate optimism with boundless energy, yet we now see a new heroic generation of climate activists and climate doers who are making the promise of Paris a reality.

Notes

1 https://www.fastcompany.com/1648639/crib-sheet-christiana-figueres-climate-change-chief-un-not-trashy-romance-novelist
2 https://www.cleaningup.live/ep6-christiana-figueres-to-paris-and-beyond/
3 https://unfccc.int/files/meetings/durban_nov_2011/application/pdf/2325_text-_9122011-indaba.pdf
4 The Bali Roadmap refers to the set of decisions reached at the 13th COP held in Bali in 2007, the main decisions being a range of measures to support developing country Parties known as the Bali Action Plan, the negotiations on the second commitment period for the Kyoto Protocol since the first commitment period was from 2008 to 2012, and the allocation of financial resources to the Financial Mechanism of the UNFCCC.

5 https://pmg.org.za/committee-meeting/13706/
6 The South African President of COP17 attempted to introduce a new multilateral process drawing upon traditional meeting practices in Southern Africa, known as Indaba, which are representative conferences.
7 https://unfccc.int/news/climate-champions-place-delivery-and-accountability-at-centre-of-post-cop26-agenda
8 https://unfccc.int/news/un-climate-summit-ban-ki-moon-final-summary
9 https://profilesofparis.com
10 https://www.theguardian.com/environment/blog/2015/dec/16/how-a-typo-nearly-derailed-the-paris-climate-deal
11 The Like Minded Developing Countries group was formed in 2012 after the adoption of the Durban Mandate in 2011 and the emergence of new alliances of progressive Parties. Its first meeting was convened by China in October 2012. The group was formed to provide a stronger voice to a traditionalist view of the UNFCCC, particularly issues related to differentiation of obligations between developed and developing countries.
12 https://unfccc.int/news/the-power-of-policy-reinforcing-the-paris-trajectory
13 Figueres, 2020

References

Figueres, C., and Rivett-Carnac, T. (2020) The Future We Choose, Manilla Press, London, UK.

IISD. (2015) Earth Negotiating Bulletin, Bonn Climate Change Conference. 19–23 October 2015, Bonn, Germany. Available online at: https://enb.iisd.org/climate/unfccc/adp2-11/

Klein, D., Carazo, P., Doyle, M., Bulmer, J., and Higham, A. (2017) The Paris Agreement on Climate Change: Analysis and Commentary. Oxford University Press, Oxford, UK.

Mission 2020. (2020) Prelude to a Great Regeneration, Imagine the Future. Available online at: https://imaginethefuture.global/resources/images/Prelude-to-a-Great-Regeneration.pdf

UNFCCC. (2011) Establishment of an Ad Hoc Working Group on the Durban Platform for Enhanced Action. Decision 1/CP.17, paragraph 8. Available online at: https://unfccc.int/files/meetings/durban_nov_2011/decisions/application/pdf/cop17_durbanplatform.pdf

8

MAURICE STRONG – FROM POVERTY TO UN UNDER-SECRETARY GENERAL AND FATHER OF SUSTAINABLE DEVELOPMENT – THE RIO EARTH SUMMIT

Felix Dodds

In 1992, world leaders, diplomats, environmental activists, and other officials descended on Rio de Janeiro for a major event—the UN Conference on Environment and Development. Dubbed the "Earth Summit," participants were hoping to launch international action on climate change, biodiversity loss, deforestation, hazardous wastes, and other threats to our global sustainability. But the issues were complex and not all countries agreed on how to proceed. Furthermore, there were others—environmental activists, scientists, business and industry groups, union and worker representatives, indigenous peoples, and many more—who wanted to be a part of the conversation and who believed their voices should be heard.

Could the Earth Summit engage with so many stakeholders successfully? Would it achieve the outcomes needed for the planet? And what was a former oil tycoon, Maurice Strong, doing at the heart of the entire event?

Our story on Maurice Strong starts in the Canadian prairie town of Oak Lake, Manitoba where he was born in 1929. Oak Lake sits on the mainline of the Canadian Pacific Railway (CPR). Maurice Strong came from a poor family. The Great Depression started the year he was born, and Maurice used to watch trainloads of the homeless and the desperate passing his home, "crossing the prairies, torn from their families by need and hunger." He later remarked that "their worn, pinched, anxious faces haunted me for years afterwards" (Strong, 2001).

Maurice would later say that those ten years of the Great Depression had radicalized him so that he thought of himself as "a socialist in ideology, and a capitalist in methodology."

His father, Frederick Milton Strong, was laid off work, like millions of others. Frederick never found a steady job after that but supported his family through odd jobs. His mother, Mary, suffered from a mental health disorder which ultimately saw her admitted to a mental institution where she died aged 56.

DOI: 10.4324/9781003202745-9

Strong was a story-book character. A self-made wealthy man with little formal education, he possessed an extraordinary gift for inter-human relations, networking and persuasion. While firmly embedded in the Western-Northern establishment he understood fully the prevailing global fractures and challenges and thus also won the trust and confidence of the South and the East. Looking beyond the immediate, he foresaw and sought to avoid problems and disasters that were bound to happen unless something was done to rise above the status quo.

(Savio, 2018)

As a child, Maurice immersed himself in nature. Nature seemed to offer more stability than the human world. He had a brother, Frank, and two sisters, Shirley and Joyce. The family survived by working for food and relying on the kindness of the community. They lived in rented properties with no central heating or indoor plumbing and wore whatever clothes they could find, handing them down to each other. His father had to go out into the bush to cut wood without proper shoes —"he'd wrap his feet in rags to keep them warm" (Strong, 2001).

Maurice dropped out of school at 14, and would never go through any formal higher education. As war was raging in Europe and the Far East, Maurice tried unsuccessfully to enlist in the war effort. He went as far as to falsify his birth certificate to say 1925 and not 1929 but was rebuffed. This did not deter Maurice, who then travelled elsewhere looking to enlist. He hitched rides on freight trains to Calgary. Ultimately, he became a stowaway on the Canadian Pacific Line steamship *Noronic*. Once the ship had left port, he announced himself available for work, and became the chief dishwasher and peeler of potatoes.

He would with his doctored birth certificate eventually enlist as a merchant marine on the *Princess Norah*, which was under contract with the US Army to transport troops to Alaska.

Maurice was an entrepreneur throughout his life. It was on the *Princess Norah* that he started to accumulate collateral by running a poker game out of his cabin, charging 50 cents out of each pot.

A year later he was hired by the Hudson's Bay Company as an apprentice fur trader. This job brought Maurice in contact with the Inuit people in the Arctic—where they dubbed him *Metik*, which means *boy-that-makes-like-a-duck*, because he did impressions of an eider duck. This experience with the Inuit influenced him throughout his life. He became a champion for Indigenous Peoples everywhere. In Canada, he eventually helped establish the Native Business Council. His wife Hanne would be a passionate advocate for the same cause in her own right.

The United Nations

The United Nations Charter was signed by 50 countries at the United Nations Conference on International Organization (UNCIO) in San Francisco in June, 1945. Maurice, even as a 16-year-old, was inspired by the vision of the United

Nations, and wanted to work for this new organization that might stop future wars and help those most in need.

The contacts he made in the extractive industries led him at an early age to Noel Monod, who had become the treasurer at the UN—a post he would hold from 1946 to 1971. It was because of Munod that Maurice became a teenage clerk in the UN Pass and ID office. Although he loved the excitement of being around the new organization, he realized that there was no way to advance within it with his educational qualifications, or rather lack of them. This deficit would require him to return to the UN through a different door, as a successful businessperson and civil servant. He told his co-workers in the Pass office that:

One day 'after I make my first million' [I will] return to lead the organization.
(Branislav, 2018)

After clerking at the UN, Maurice took time off to travel the world as a 22-year-old. He booked passage on a ship going through the Suez Canal to South Africa. While stopping off at Mombasa, he heard there were jobs in Nairobi. There he found work with an oil company, Caltex, which enabled him to travel extensively in East Africa. While there, he saw the terrible impacts of colonialism. He was sickened by the actions of the British colonial regime in Kenya, which would continue until Kenya's independence on 12 December 1963.

Maurice spent most of his early working life in one or another job with oil and mining companies—scarcely the typical place for an environmentalist to begin his working life.

The Vincent Mining Corporation took Maurice on as an accountant trainee. This would lead to several jobs including one with Norcen Resources, a small natural gas company which he would run to become one of the largest in the industry. During his early years in industry, he made political contacts for later work on the international development and environmental issues that he really cared about.

With his understanding of the sector, it is hardly surprising in 1976 that Prime Minister Pierre Trudeau would turn to him to set up and direct Petro-Canada.

Maurice would be called many things. One description that he remembered in his 2001 autobiography always made him smile: "anti-charismatic to a fault, the Michelangelo of networking, a would be king of the world" (Strong, 2001).

Maurice had through the 1960s been on many corporate boards and had taken an increasing interest in foreign affairs and development issues, in particular.

Prime Minister Lester Pearson recruited him into the Canadian government. Lester Pearson's Liberal government had introduced universal health care and the Canadian Pension Plan, but it was love for the UN that they both shared that brought Pearson and Strong together. In 1957 Pearson won the Nobel Peace Prize for coordinating the UN Emergency Force sent to solve the Suez Canal Crisis. Pearson asked Maurice to come into the government as a deputy minister for External Aid and to establish the Canadian International Development

Agency (CIDA). This position at last brought him to UN meetings, where he quickly became well known for good advice.

Preparation for the 1972 Stockholm UN Conference on Human Environment wasn't going well, so UN Secretary General U Thant, impressed with what Maurice had achieved at CIDA, sought him out to be Secretary General of that Conference. At the age of 42, Maurice became the first Under Secretary General for the Environment. He would hold the post of Under Secretary General of the United Nations an amazing seven times in his career.

The Stockholm conference had 114 governments attending and its outcome was a Declaration of Principles to guide interaction with the environment and an Action Plan to execute the work. It is remembered for many things which also have Maurice's fingerprints on them, including the Club of Rome's 'Limits to Growth Report' and the 'Founex Report on Development and Environment', which helped bring development experts and countries onboard as an alternative to boycotting the conference. Stockholm was the first UN Conference attended by the People's Republic of China which on 25 October 1971, had replaced the Republic of China (Taiwan). It would lead to China taking environmental issues more seriously than some countries.

The Stockholm conference also generated the first UN text recognizing that there might be a problem with climate change:

> Recommendation 70: It is recommended that Governments be mindful of activities in which there is an appreciable risk of effects on climate, and to this end:
>
> (a) Carefully evaluate the likelihood and magnitude of climatic effects and disseminate their findings to the maximum extent feasible before embarking on such activities.
> (b) Consult fully other interested States when activities carrying a risk of such effects are being contemplated or implemented.
>
> *(UN, 1972)*

After the Stockholm Conference, Maurice was asked to establish the first intergovernmental body to focus on the environment. As governments across the world started to create ministries of the environment, it was his job to create the United Nations Environment Programme (UNEP) which he then directed for nearly three years. There had been an agreement in the UN that after the creation of UNIDO in Vienna, the next UN body created would reside in a developing country. Maurice found himself Nairobi again, working with the government of Kenya. UNEP was the first UN institution established in a developing country.

After the Stockholm Conference, global environmental challenges came to the fore. These included acid rain, deforestation, air and water pollution, the destruction of the ozone layer, and early signs that governments needed to address climate change. UNEP was at the centre of facilitating science for these challenges and, where appropriate, multilateral environmental agreements (MEAs) to address these problems.

The Run Up to the Rio Earth Summit

The most logical secretariat for the 1992 United Nations Conference on Environment and Development (UNCED), otherwise known as the Earth Summit, would have been UNEP. The obvious Secretary General of the conference should have been Mostafa Tolba, then the Executive Director of UNEP.

Tolba had led the Egyptian delegation to the 1972 Conference on Human Environment and been Maurice's deputy when UNEP was established. Maurice left UNEP in 1975 and returned to Canada, with Tolba taking over as UNEP's second Executive Director. Tolba established UNEP as the world's voice for the environment over the next 17 years. He helped UNEP build scientific consensus on key environmental challenges, which led to the creation of many key MEAs.

During the time he was head of UNEP, Tolba had cajoled or bullied member states to take up key environmental issues. This included the hugely successful Vienna Convention on Ozone Depleting Chemicals and its critical Montreal Protocol.

Tolba had also played a significant role in preparing the ground for what would become the UN Framework Convention on Climate Change through the two World Climate Conferences (1979, 1990) and the UN Convention on Biological Diversity.

The Canadian government approached Maurice in the summer of 1989 to see if they could put his name forward to be Secretary General of the UNCED—this had surprised Maurice as he wasn't close to the conservative Prime Minister Brian Mulroney (1984–1993).

Member states wanted a "fresher" pair of hands but also someone with experience and so the idea of Maurice Strong was appealing. Afterall, he had established UNEP and been Secretary General to the first UN conference on the Environment in 1972. Nostalgia perhaps also played a role in the choice but also in those years since Stockholm, Maurice had continued to play a leading role in promoting environmental issues and sustainable development.

A UN General Assembly resolution in 1989 established the involvement of stakeholders and an independent secretariat for the conference process. It called on different UN Agencies and Programmes to work with the conference secretariat to ensure that the full UN system would contribute to the UNCED. To ensure this cooperation, the Secretary General, anticipating competition between parts of the UN, gave the Administrative Committee on Coordination (ACC) a role in ensuring a supportive approach by the UN system. The ACC was the closest thing to a cabinet of the UN Agencies and Programmes that existed then.

Having Maurice back and an independent secretariat for the conference established a good foundation for what was going to happen and ensured fewer inter-agency conflicts.

By 1990 he environmental agenda and development had grown and become more urgent after the Stockholm Conference in part because governments had not implemented much of the 1972 agreements.

Maurice wondered whether outcomes would have been different if he had done things differently at Stockholm. The Rio Earth Summit offered him the opportunity to try some new ideas.

The UN had appointed Maurice with a very limited budget, and a staff of only 22. He recognized that would not be nearly enough to support what he had in mind, which included bringing in staff from outside the UN. He wanted an additional budget to supplement his own staff with people he gathered through other means. He turned for help to Benjamin Read, who had been the first President of the German Marshall Fund. Together they established a non-profit foundation called the Eco-Fund to mobilize private financing. This move ultimately enabled Maurice to have a staff of 99 and a network of unpaid advisers that included key stakeholder leaders. The UN opened a formal investigation into whether he could do that and found that there was nothing to stop him from doing it.

The History of the Stakeholder Engagement

In 1982 in preparation for the ten-year anniversary of the Stockholm Conference on Human Environment) the International Union for Conservation of Nature (IUCN) produced the 'World Conservation Strategy: Living Resource Conservation for Sustainable Development'. The Strategy is most remembered as the first report to coin the term "sustainable development" in an attempt to reconcile economic development with environmental protection (IUCN, 1980).

IUCN is an interesting example of an early multistakeholder organization which had governments and stakeholders, particularly scientists, in its membership.

That need to address economic development and the environment had already begun at the Stockholm Conference, but by the 1980s the need to address this was becoming urgent.

The Secretary General of the United Nations, Javier Pérez de Cuéllar, asked former Norwegian Prime Minister Gro Harlem Brundtland to head a 'World Commission on Environment and Development', renamed the 'Brundtland Commission' by the media. Maurice was one of the members of the Commission, and in its report 'Our Common Future' defined sustainable development for the first time in a UN document:

> Sustainable development is development that meets the needs of the present without compromising the ability of future generations to meet their own needs.

> *(UN, 1987)*

It not only brought together environment and development but also the recognition of the responsibility of this generation for future generations—a recognition of inter-generation equity.

The UNCED was a direct response to one of the Brundtland Commission's recommendations:

> Special follow-up conferences could be initiated at the regional level. Within an appropriate after period after the presentation of this report to the General Assembly, an international conference could be convened to review progress made, and to promote follow up arrangements that will be needed to set benchmarks and to maintain human progress.
>
> *(UN, 1987)*

The Rio Earth Summit Process

The decision by the UN General Assembly to host the UNCED (UN, 1989) laid out the timetable for the conference process as four two-week preparation committee meetings (PrepCom) in Nairobi (1), Geneva (2 and 3), and New York (4) with the conference in Rio for two weeks to coincide with World Environment Day, 5 June 1992.

The final PrepCom ultimately spanned not two, but *five* weeks, and the preparatory meetings totalled 98 days, which is still a record for a UN Conference or Summit process.

Eastern bloc countries had boycotted the Stockholm Conference because East Germany was not allowed to attend as it wasn't recognized by any UN body. By the time of the Rio Conference, former Soviet satellite countries were independent democratic countries, and Rio was the first major UN Summit or Conference to be held in the post-Soviet Union era of history.

The hoped outcome of the Rio Earth Summit was Agenda 21, a 40-Chapter 'Blueprint for the 21st Century', three legally binding conventions on Biodiversity, Climate Change, and Forests, and an Earth Charter that described ethical principles for living more sustainably and peacefully. By the end of the Earth Summit what was achieved was Agenda 21, two conventions on Biodiversity and Climate Change, a set of Forest Principles and not an Earth Charter but a much less inspiring but still important set of principles to guide governments known as the Rio Declaration. (Parson et al., 1992)

This chapter cannot take any of these outputs and map out the role that Maurice played using his political sensitivity and skill to steer huge agendas to a successful conclusion. Our focus, instead is on one of the great innovations that Maurice brought during the Rio process, which would change the way the UN would interact with stakeholders forever. It was the Earth Summit in Rio in 1992 changed the understanding of the term "NGO" to "stakeholder" within the UN, expanding stakeholders' role in policy development.

Agenda 21 had nine chapters (of the 40 total) dedicated to the rights and responsibilities of stakeholders, which it called Major Groups. These nine were: Women, Children and Youth, Indigenous Peoples, Non-Governmental Organizations, Local Authorities, Workers and Trade Unions, Business and Industry, Scientific and Technological Community, and Farmers[1].

And So, It Starts—Nairobi August 1990

From UNEP headquarters in Gigiri, Kenya, just outside Nairobi, Maurice organized the first preparatory meeting for what would become the known as the Rio Earth Summit.

He had many things on his mind.

The Stockholm commitments had not been implemented satisfactorily. Maurice resolved to not rely on governments alone to deliver outcomes. He had been inspired by the energy and commitment of NGOs in the NGO Forum in the 'Farm' during the Stockholm conference, but the conference hadn't allowed their voices to have the impact they should have had.

An early decision by this PrepCom to allow broader engagement with stakeholders would potentially help bring thousands rather than hundreds of stakeholders to the conference. If they came but couldn't access the negotiations or organized events, the process would be called greenwash. The challenges facing the world were too great to let that happen.

Maurice chose Nitin Desai as his deputy. He had worked with Nitin, an Indian government economist, on the Brundtland Commission. Both had been impressed by the engagement of stakeholders in the Town Hall meetings that the Brundtland Commission had hosted all over the world. The one that impacted the final report most significantly had been held in São Paulo, Brazil. Many people from community groups described links between environment and development there.

Member states elected Tommy Koh, the Singapore Ambassador to the United Nations, as the chair of the Bureau for the Summit. Between Desai and Koh, Maurice had great partners for the journey that was about to happen.

One of the items high on their list of points to do was to expand the engagement of NGOs. This was the UN term for groups which are not governments. Stockholm had shown Maurice that they needed to have greater access to negotiations if they were going to help governments make better decisions. There also needed to be different NGOs than those that already had consultative status to the UN. The NGO group needed to include those engaged at the national and community levels who would bring knowledge of good and bad policies to the process. After all, they were at the sharp end of any policy implementation. But just adding more NGOs wouldn't be enough. Maurice recognized that he needed them grouped into different constituencies, or as we now call them, stakeholder groups. He needed to change the rules or break them.

Fortunately, PrepCom 1 was away from New York and the more conservative member state mission delegates.

No one took much notice of what was being put together in this first Prep-Com, but that would change as governments engaged at higher levels through the subsequent PrepComs.

Tommy Koh's gavel echoed in the UNEP Conference room to start the process. It would be a tool he used to good effect throughout the planning.

By the end of that first PrepCom, sitting next to Tommy, Maurice looked across the room and smiled. His team succeeded in expanding the engagement of stakeholders beyond those already consulting with the UN. Now organizations could accredit to the conference process, and this would enable a wider level that could bring real experiences to the delegates. Moreover, stakeholders could speak in the formal sessions of the conference.

The drafting of the chapters for Agenda 21 would start shortly and Maurice was keen this wouldn't be undertaken by the UN by itself but would be supported by the input from experts from different stakeholder groups.

Five Weeks in New York, March 1992

Now that PrepCom 1 had sorted out opening the process, Maurice visited as many countries as he could to build up key constituencies. In every country he visited, his advance teams set up three meetings: with the government minister and/or Head of State, with the media, and with stakeholders. He began to recognize that the final agreement would need text tied to delivering the outcome projected by Agenda 21. The mantra Maurice repeated to himself was, "We can't trust governments by themselves to do what they say they will; We need to bring other stakeholders into the process."

One way to understand what stakeholders could contribute was to appoint a group of stakeholder advisors. This group helped the secretariat with outreach but also in crafting a substantive commitment to become implementation agents for Agenda 21. This insight created constituencies to monitor governments on following up with their commitments.

Maurice turned to another former Brundtland Commission staff person, Chip Lindner, who had been secretary to the Commission and had set up *The Centre for Our Common Future* to promote outcomes from the Commission's report. Linder brought together stakeholders, or as the Centre called them, the independent sector. He also understood the wisdom of expanding beyond the big NGOs and the NGO stakeholder group.

In the period between that first PrepCom and the fourth and final PrepCom in New York starting on 3 March 1992, much had happened, most of it in response to the open-door approach that Maurice championed. Many stakeholder groups had held their own conferences. In some cases, new institutions had been set up to bring voices to the table. This included the International Council for

Local Environmental Initiatives led by Jeb Brugmann, and the Business Council for Sustainable Development set up by Swiss industrialist, Stephan Schmidheiny.

Over 1,000 stakeholders attended the final PrepCom, and it was here that Maurice tabled the first secretariat text with the nine chapters on the roles and responsibilities of different stakeholder groups (Major Groups). Maurice and Tommy had noted which stakeholders had been active and mentioned by member states, so that what the secretariat played back was something that member states had already heard in one way or another.

Maurice had prepared the ground for the final PrepCom by meeting with Ministers and delegates over the previous 18 months. Many had seen growing interest in the Summit at home because conferences were being held, and pressure was put on Heads of State to attend the Summit. Added to that, many good ideas from the stakeholders enriched the text.

Secretariat drafts were circulated under a new section of Agenda 21 called 'Strengthening the Role of Major Groups'.

Tommy Koh asked Bureau member *Ambassador Leon Mazairac* of the Netherlands to lead the sessions on the Major Groups. Mazairac built on the work of Maurice and Tommy and called on them for consultations with member states to underline the support of the Bureau and Secretariat.

Tommy Koh opened PrepCom 4, the first PrepCom in New York, where some of the more conservative delegates were focusing on the proposed text for the first time. Some questioned the new level of stakeholder engagement achieved in the balmy August temperatures of 1990 in Gigiri. Koh reminded member states of the decision their colleagues had taken. Koh explained that the Bureau had agreed that stakeholders would be given access to both formal and informal process. They couldn't speak in the informal sessions, but the chairs of the session could go to a formal session to speak—something never done before.

PrepCom 4 also was the first place where stakeholders could book UN rooms for their meetings, helped by the seemingly inexhaustible Yolanda Kakabadse, the Secretariat NGO Liaison. Many stakeholders were frustrated by not having current information about what was going on in the informal meetings or contact groups that they were not allowed to attend. Koh in response to this asked the chairs of the Working Groups to make themselves available to brief the stakeholders each day. This would be something repeated throughout subsequent UN processes, but this is getting ahead of ourselves.

PrepCom 4 opened to a full gallery of stakeholders and a full room of member states. It wasn't until the afternoon of 16 March, that the plenary saw the first discussion on the Major Groups section, and several governments supported adding a section on youth to those proposed by the secretariat. They also considered adding a section on the elderly, which unfortunately did not make it into the final text.

Before the next plenary on 23 March, there were informal consultations with representatives from the different stakeholders' groups. Maurice was impressed by the passion that progressive women had shown throughout the process. They were led by former US Congresswoman Bella Abzug, a force of nature whom some people called a hurricane. Informal consultations were also held with key governments to seek them as strategic advocates.

This Plenary was opened not by *Ambassador Mazairac* but by Tommy Koh, with Maurice sitting next to him again to underscore support for these chapters in the Bureau and Secretariat. The session moved to opening statements by representatives of four of the major groups: Vandana Shiva (women), Flavia Ferreira (youth), Evaristo Nugkuag (indigenous people), and Ramsing Hurrynag (NGOs). Other stakeholders would be heard at the next session. After the four presentations, Tommy handed the gavel to Ambassador Mazairac. The session cleared the text on women in one session with Bella present in the room wearing one of her iconic hats, and many delegates who had felt her wrath didn't want to feel it again.

Norway had been a voice for the amendments that the women's groups wanted. India tried unsuccessfully to reduce the commitment in the relevant chapters about the role of governments from "should" to "could", which would have gutted the chapters. The meeting adjourned with 116 paragraphs yet to be agreed on, and only one more scheduled session planned.

On 24 March, the session started with the chapter on youth. No real progress was made, and the session would reconvene the next day. At that session, Sweden, a strong supporter of the 'Convention on the Rights of Children' (signed on 20 November 1989) successfully proposed adding children to the chapter name and intent, which would subsequently be called 'Strengthening the role of children and youth'.

Initial statements were offered on the chapters about indigenous peoples, NGOs, local authorities, trade unions, business and industry, the scientific and technological community, and farmers.

Ambassador Mazairac moved from general comments to negotiating article by article. Where problems arose, he established contact groups to solve them.

Tommy and Maurice worked behind the scenes as needed, meeting with problematic member states. They would calm their worries, explaining the process but also reminding all concerned how much stakeholders had already contributed to making the document better. If stakeholders could do that, couldn't they also help in implementing it?

By PrepCom 4 something like 15% of governments had stakeholders on to their delegations, many for the first time (Pallas, 2016).

Conversation by conversation, coffee by coffee, nudge by nudge, even the most challenging countries turned to amending the text as opposed to opposing it, and so the text slowly became more of a member state text, with arguments focused on what should be included or excluded.

By March 26, the children and youth chapter was mostly clear text and ready for approval, and a contact group would deal with a few outstanding issues. Canada led the push for the chapter, and a youth delegate on their delegation often proposed amendments to the text.

France expressed reservations about any text that might strengthen policies to support indigenous peoples. Malaysia worried that the text about the role of NGOs blurred the distinction between what NGOs and governments were supposed to do.

On March 28, the member states met all day in contact groups and cleared the text of five other stakeholder groups, reaching consensus on the roles of:

1 indigenous people and their communities.
2 non-governmental organizations.
3 local initiatives in support of Agenda 21.
4 workers and their trade unions.
5 business and industry.

That left only the roles of scientists and farmers still to be addressed.

At 3:45 am on Saturday, 4 April, the Plenary turned its attention to the last section of Agenda 21 it had to adopt, 'Strengthening the role of Major Groups'.

A youth delegate opened with a critical statement about where the negotiations had ended and then debate started.

France lifted the reservations it had placed on the Indigenous Peoples text. Malaysia also withdrew its reservations. Kenya suggested that what they had in front of them should be adopted, and then left to Ambassador Mazairac to continue consultations on any remaining problems.

After the document was adopted, but before everyone could congratulate themselves and think of sleeping again, the Palestinian representative wanted any reference to Major Groups to include 'People under Occupation'. This demand resulted in Chapter 18, *Protection of the quality and supply of freshwater resources: application of integrated approaches to the development, management and use of water resources* adding 'people under occupation' to a list of major groups that should be involved in water resources management. Israel controls all of the water going into areas that the Palestinian representative would like to see under a Palestinian government. While this would be an issue at the Earth Summit two months later, most delegates just wanted to sleep.

Was Maurice relaxing after the fourth PrepCom and reflecting on what had been achieved? Something like 85% of Agenda 21 was agreed to, including nearly all of the text on the nine chapters on stakeholders. He couldn't resist a tired smile. He knew where the problems were on the remaining text. He, Nitin, and Tommy had work to do before member states would reconvene on 3 June in Rio de Janeiro.

The Earth Summit

When Maurice delivered his opening speech at the Earth Summit, representatives from 172 nations were in front of him, with 120 Heads of State attending at RioCenro, the conference site. This was the largest number ever attending such an event. There were also 2,400 representatives from different stakeholders, 10,000 journalists, and a further 17,000 stakeholders who attended the parallel Global Forum occupying green and white tents in Flamenco Park in Rio, 40 kilometres away.

The Global Forum was a "world fair of environmentalism" which would inspire many of those who attended to take Agenda 21 back to their communities across the world.

Near the end of his speech, Maurice turned to the important role of stakeholder constituencies as critical to the implementation of Agenda 21:

> No constituencies are more important in all countries than women, youth and children. And the children who greeted us so beautifully at the door as we entered today - representatives of all the children whose world we are shaping here must surely be a particularly poignant reminder of the special responsibilities we carry towards them. To make their essential and distinctive contributions, the remaining barriers to the full and equal participation of women in all aspects of our economic, social and political life must be removed. Similarly, the views, concerns and the interests of our youth and children must be respected and they must be provided with expanding opportunities to participate in the decisions which will: shape the future which is so largely theirs.
>
> By the early part of the 21st century, more than half the world's people will live in urban areas. Cities of the developing world are being overwhelmed by explosive growth at rates beyond anything ever experienced before. By the year 2025, the urban population of developing countries is expected to reach some 4 billion. In our host country, the proportion of people living in urban areas is already more than 70%. The meetings of leading representatives of local governments, in which Mayor Alencar took such a leading role, which took place in Curitiba and Rio in the past week, have highlighted these issues and established the basis for the adoption of an Agenda 21 by many of the world's leading cities.
>
> "We are reminded by the Declaration of the Sacred Earth Gathering, which met here last weekend, that the changes in behaviour and direction called for here must be rooted in our deepest spiritual, moral and ethical values. We reinstate in our lives the ethic of love and respect for the Earth which traditional peoples have retained as central to their value systems. This must be accompanied by a revitalization of the values common to all of our principal religious and philosophical traditions. Caring, sharing, co-operation with and love of each other must no longer be seen as pious ideals, divorced from reality, but rather as the indispensable basis for the new realities on which our survival and well-being must be premised.
>
> Science and technology have produced our knowledge-based civilization. Its misuse and unintended effects have given rise to the risks and imbalances which now threaten us. At the same time, it offers the insights

we need to guide our decisions and the tools we need to take the actions that will shape our common future. The guidance which science provides will seldom be so precise as to remove all uncertainty. In matters affecting our survival, we cannot afford to wait for the certainty which only a post-mortem could provide. We must act on the precautionary principle guided by the best evidence available."

(Strong, 1992)

The Summit was also the first UN event to see the internet used to collaborate and to put pressure on governments. Though in its infancy stakeholders found ways to ensure that their governments were aware of what was being advocated in capitals. The use of email to fax proved to be a useful tool so that overnight a hotel might collect the faxes that had come in and give them to the government officials, perhaps as they sat down for breakfast.

When discussion about the Major Group section of Agenda 21 opened on 5 June, a number of paragraphs in all of the chapters remained in brackets. Ambassador Mazairac asked for further consultations. Member states reconvened on 11 June with clean text for Chapters 23 (preamble), 24 (women), 25 (children and youth), 28 (local authorities), and 29 (workers and trade unions). The Ambassador put forward compromise text which addressed the remaining bracketed paragraphs for Chapters 26 (indigenous people), 27 (NGOs), 30 (business and industry), and 32 (farmers). The relevant paragraphs were quickly adopted.

This involved agreement on eight of the nine major Group chapters. The controversy over adding 'Occupied Territories' language had been resolved, and that left just one chapter still to be sorted out.

Surprisingly it was Chapter 31 on the scientific and technological community. Maurice, knowing that there had been problems in that chapter in PrepCom 4, had used his opening speech to underline the importance of science and technology in his final comments opening the conference:

To become full partners in the process of saving our planet, developing countries need first and foremost substantial new support for strengthening their own scientific, technological, professional, educational and related institutional capacities. This is one of the important and urgent features of Agenda 21.

(Strong, 1992)

So where were the problems coming from?

The Holy See and their supporters in the member states wanted to see the phrase 'appropriate ethical principles' included. There is no question that the present Pope, Francis, accepts the need to address climate change. But the Earth Summit was held in 1992, in the middle of the reign of Pope John Paul II, one of the more doctrinally conservative Popes.

Tommy Koh, the Chair of the conference, prevailed upon member states in the Committee to adopt the text with his suggestion to note the Holy See's concerns.

Engaging with the World

One of the lasting legacies of the Earth Summit was the engagement of stake-holders in the process to such an extent that nine of the chapters of Agenda 21[1] would identify the roles and responsibilities of stakeholders. Agenda 21 would be the first UN text to recognize the theory of change:

> Involving stakeholders in the decision making makes better-informed de-cisions and that those decisions are more likely to be implemented with those stakeholders' support either singularly or in partnership.
>
> *(Dodds, 2019)*

It would take another eight years for the UN Global Compact to be created and the UNEP Finance Initiative started aligning the finance sector to address sustainable development. Maurice would have been very happy in 2021 to see the launch of the Glasgow Financial Alliance for Net Zero (GFANZ), chaired by Mark Carney, UN Special Envoy on Climate Action and Finance, which unites over 160 firms (together responsible for assets in excess of US$70 trillion) across the financial system to accelerate the transition to net zero emissions by 2050 at the latest (Race to Zero, 2021).

The Earth Summit in 1992 was the starting place for engagement of industry to address environmental and sustainable development challenges. It was the first place where local or sub-national governments really got a seat at the table to play their role in moving communities towards a more sustainable path. It was the birthplace of much of the intergovernmental framework that now directs the work on sustainable development. It also inspired a generation to engage in sustainable development, and they then inspired their children.

Post the Earth Summit

After Rio, Maurice continued working in industry (WBCSD, 2000) and with non-profits, advising governments, and returning for the last time as an Under Secretary General to help UN Secretary General Kofi Annan with UN reform. He played a supporting role for the Rio+20 (2012) conference. As was always his way, Maurice highlighted issues that needed to be addressed by Rio+20 and contributed to the development of the Sustainable Development Goals (SDGs), which were in many respects daughters of Agenda 21.

Earth Council and Charter

The hope of Maurice had been to have an Earth Charter as an outcome from the Summit but that turned out to not be possible, so it's (more modest) replacement was the Rio Declaration.

To take forward the idea of the Earth Charter and other key issues that needed more work, Maurice established the Earth Council. Its work included helping indigenous peoples relate their spiritual traditions and value systems to the movement for sustainable development and creating: "an 'ombudsman' function to help redress environmental injustices and resolve disputes" (Strong, 2001).

The Council was the first NGO to be mentioned in a UN Conference outcome document, even before it was formally constituted (Maurice was a master of networking).

> 38.45. The Conference takes note of other institutional initiatives for the implementation of Agenda 21, such as the proposal to establish a non-governmental Earth Council and the proposal to appoint a guardian for future generations, as well as other initiatives taken by local governments and business sectors.
>
> *(UN, 1992b)*

The Earth Council was established in Costa Rica in 1992 and played an important role in helping governments to establish over 80 National Councils for Sustainable Development in its first five years. It also hosted the controversial 1997 Rio+5 Conference, which leading NGOs saw as a distraction from a proper assessment of governments' delivery on Agenda 21.

Mikhail Gorbachev's Green Cross International, established in 1994, joined with Maurice in supporting the idea of an Earth Charter. It took six years of consultation with stakeholders to develop the final Charter, which debuted in 2000 at The Hague. There was an expectation that it might be recognized at the World Summit on Sustainable Development (2002) and it appeared in the first draft of the political declaration there, but not in any further drafts. It was registered as a multi-stakeholder partnership at the Summit, but this did not require political endorsements. The Earth Charter was created to:

> Establish a sound ethical foundation for the emerging global society and to help build a sustainable world based on respect for nature, universal human rights, economic justice, and a culture of peace.
>
> *(Earth Charter Secretariat, 2000)*

The Charter won recognition in 2003 with UNESCO, but since then it has not succeeded in being adopted by other intergovernmental organizations. When Pope Francis issued his climate encyclical, he praised the Earth Charter as a document that "Asked us to leave behind a period of self-destruction and make a new start" (Pope Francis, 2015).

UN Commission on Sustainable Development

To monitor the delivery of Agenda 21 a new subsidiary Commission to the UNE-SCO was established, the Commission on Sustainable Development (1993–2012).

As Agenda 21 had identified nine stakeholders that were critical to the delivery of the outcome document, it had to decide how to operationalize that.

The idea of involving stakeholders was a groundbreaking decision. As mentioned earlier prior to this all those who were not national governments were grouped at the UN by the term "non-governmental". This solidified the recognition that there was a need for stakeholders to advise government decision-making process and also to be involved in the implementation. This would carry forward in the coming decade in all the major sustainable development processes at the UN. Maurice had been key to this because he recognized that involving stakeholders in the decision-making process helped governments to make better informed decisions.

Maurice also wanted many of the stakeholders who had participated in the 1992 Conference to become accredited by the UN to participate in the Commission on Sustainable Development. Using his political capital after the Summit, he pushed for those stakeholders who had been at Rio to be asked to see if they wanted UN accreditation. The first step was the ECOSOC decision on 4 March 1993 that any NGO accredited to the UNCED "could apply for and should be granted Roster status" (ECOSOC decision 1993/215).

A total of 550 stakeholders said that they would pursue accreditation and were registered (UN document E/1993/65) to attend the first CSD. With the help of the first Chair of the CSD Ambassador Razali Ismail of Malaysia, Maurice then pushed for them to be admitted as formally accredited to the UN. This was eventually agreed—despite great reluctance from countries that did not want NGOs at the UN in the first place—with ECOSOC decisions 1996/302. The number of NGOs that were in the end added as CSD Roster stakeholders was 539. The definition of being on the Roster was that the organization was small or highly specialist or worked with a particular UN Agency. This influx of new organizations to the UN changed the organization's work completely over time.

In the 1990s and early 2000s the CSD became the test bed for new arrangements with stakeholders. The hosting of stakeholder meetings pioneered in the Earth Summit process was solidified and spread across the UN. Speaking in formal and even informal sessions – at the discretion of the chair – became standard, and to the annoyance of some member states, even found its way into the UN General Assembly process around Rio+5 helped by the fact that Ambassador Razali was now President of the UN General Assembly and used his weight to operate by "CSD rules of stakeholder engagement." The CSD introduced first a day on different stakeholders, and then in 1996 for the five-year review and subsequent CSDs, two days of dialogue with stakeholders. The legacy of those first ten years is huge, not only on how the UN interacts with stakeholders but also on how member states are doing so at the national level.

Slowly the sustainable development world started to follow suit, and instead of just treating NGOs as a term for everyone who wasn't government, they used the nine stakeholder (Major Groups) chapters. By 2022 the successor to the CSD, the High-Level Political Forum, the United Nations Environment Assembly, the Rio Conventions, the FAO Committee on Food Security, and sessions of the Economic and Social Council and the UN General Assembly all wanted to hear individual stakeholder views.

The Agenda 21 process had an impact on national policy making with over 100 governments setting up national commissions or councils on Sustainable Development. At the local level, by 2002 over 6,000 authorities had developed with their communities a 'Local Agenda 21'. These efforts then influenced the way that sustainable development was being integrated into local policy making.

UN Reform

There had been many attempts to reform the UN since its creation, with differing levels of success. It is not surprising that when Kofi Annan took his post as UN Secretary General (1997) he turned to Maurice and asked him to be UN Executive Coordinator of Reform. Maurice with a small but dedicated staff took six months to produce the report 'Renewing the United Nations: A Programme for Reform'. These reforms included bringing a sprawling UN secretariat into a much more focused set of departments.

University of Peace

The United Nations General Assembly had established the University of Peace (UN, 1980) and in 1999 at the request of UN Secretary-General Kofi Annan Maurice would become its Rector until 2001. He had been sent in to protect the UN's reputation, because the organization had suffered from mismanagement, misappropriation of funds, and inoperative governance (Strong, 2004).

The University of Peace was located in Costa Rica. It focused on studies and training in the area of peace, peace for education, and human rights. During his time at the University, Maurice helped to revitalize the institution and rebuild its academic program. It has the unique ability to award degrees that are recognized by all the members of the UN General Assembly (University of Peace, 2018).

China

Maurice had built a good relationship with China since 1972. He recognized that as they developed, they would play a critical role politically and would need to ensure that development was sustainable. He spent a lot of time in China as an advisor after the 2002 World Summit on Sustainable Development helping them start to take a more active role in sustainable development.

Climate Change

Maurice believed that climate change would be the defining issue of this generation.

No issue is more important to the human future than that of climate change in which the political will to act cooperatively and decisively has dangerously diminished.

Rio+20 must reinforce international efforts at Durban and beyond to reach agreement and renewal of the Climate Change Convention and its implementation. Paradoxically, if we fail to act, the reduction in global greenhouse gas emissions could occur through the collapse of the world economy, to which none of us would aspire. After all, the roots of the environmental and climate change crises are the same as those of the economic and financial crises – the inadequacies of our economic system.

(Strong, 2011)

Maurice passed away on 27 November 2015, so could not be in Paris when the Paris Climate Agreement was secured. There is no question that without the commitment that Maurice gave to addressing the challenges that climate change would pose for humanity from 1972 to the time of his passing, the Paris Climate Agreement would have been what it was. Maurice promoted the issue over nearly 45 years, nudging, pushing, or pulling governments to address it.

A Final Reflection on the Key Outcomes from the Earth Summit

The Earth Summit ultimately birthed the largest number of legally binding MEAs ever seen.

In addition to the conventions on climate change and biological diversity in 1994, a third convention to Combat Desertification would be agreed to. In 1995, the United Nations Fish Stock Agreement codified optimum utilization of fisheries resources both within and beyond the exclusive economic zone. Out of Agenda 21 *Chapter 19 Environmentally Sound Management of Toxic Chemicals, Including Prevention of Illegal International Traffic in Toxic and Dangerous Products* emerged the negotiations that called for "a legally binding instrument on the voluntary PIC procedure by the year 2000" (UN, 1992c). This resulted in the Rotterdam Convention on the Prior Informed Consent Procedure for Certain Hazardous Chemicals and Pesticides in International Trade (1998).

Maurice helped to create much of the international framework for sustainable development that we take for granted, such as the UNEP, UN Habitat, the Intergovernmental Panel on Climate Change, the UN Framework Convention on Climate Change, the CSD, and reforms in other key institutions.

It wasn't just at the UN that he had a huge impact; within the Canadian government, Maurice helped to create the CIDA.

In the stakeholder world, he played an important role in the establishment of organizations such as the International Institute for Environment and Development, the International Institute for Sustainable Development, the Earth Council, the World Business Council, the International Council for Local Environmental Initiatives.

Without the Rio Earth Summit there may never have been momentum behind the climate change agenda, advancement of stakeholder engagement, and their impact at the UN and other intergovernmental bodies. There probably would not have been the UN Global Compact and Rio+20 which created the process for the SDGs. The SDGs are a more focused version of Agenda 21. They build on the work undertaken by those who tried to implement Agenda 21.

Maurice was not universally loved. He made the right wing in the United States go crazy. He told the silliest jokes, he could be insecure and curt to you if he thought you were doing something against what he thought was best. But he could also be truly kind. He mentored many people who would go on to hold important positions in the intergovernmental, governmental, industry, NGOs, and the larger stakeholder world. (Schwab, 2018)

All of us active on these issues today sit on the "shoulders of a giant"; it is unlikely that someone like a Maurice Strong will come along in the intergovernmental world again. (Desai, 2018) Perhaps the closest analogue outside that world is Elon Musk or Bill Gates in industry.

A final reflection on Maurice from former UN Secretary General Kofi Annan:

> If the world succeeds in making a transition to truly sustainable development, all of us will owe no small debt of gratitude to Maurice Strong, whose prescience and dynamic presence on the international stage have played a key role in convincing governments and grassroots alike to embrace the principle – if not the practice – of adopting a new, long-term approach to the global environment.
>
> *(Annan, 2001)*

Note

1 Major Groups Agenda 21 (Chapters 25–32): Children and youth, indigenous people and their communities, non-governmental organizations, local authorities, workers and their trade unions, business and industry, scientific and technological community, and farmers.

References

Annan, K. (2001) Introduction to 'Where on Earth Are We Going?' by Maurice Strong. Texere.

Branislav, G. (2018) Maurice F. Strong: A Historic Figure in the UJ and in the North-South Development Saga in Tributes and Reminiscences edited by Frederico Mayor, Negoslav Ostojic and Roberto Savio, European Centre for Peace and Development, University of Peace.

Desai, N. (2018) Maurice Strong – A Global Citizen in Remembering Maurice F. Strong: Tributes and Reminiscences edited by Frederico Mayor, Negoslav Ostojic and Roberto Savio, European Centre for Peace and Development, University of Peace.

Dodds, F. (2019) Stakeholder Democracy: Represented Democracy in a Time of Fear, Routledge.

Earth Charter Secretariat. (2000) Earth Charter Initiative: Brief History, Earth Council. Available online at: https://webpages.scu.edu/ftp/kwarner/8-84-EarthCharter.pdf

Establishment of the University for Peace, UN. Available online at: http://ecpd.org.rs/pdf/agree/Establishment_of_the_University_for_Peace.pdf

International Conservation of Nature. (1980) World Conservation Strategy: Living Resource Conservation for Sustainable Development, IUCN.

Pallas, J. (2016) UN Environment Conferences and Civil Society Participation, Young Australians in International Affairs. Available online at: https://www.youngausint.org.au/post/2016/10/23/un-environment-conferences-and-civil-society-participation

Parson, E., Hass, M., P., and Levy, A.M.A. (1992) Summary of the Major Documents Signed at the Earth Summit and the Global Forum, CIESIN. Available online at: http://ciesin.org/docs/003-312/003-312.html

Pope Francis. (2015) Encyclical Letter Laudato Si' of the Holy Father Francis on Care for Our Common Home, Vatican. Available online at: https://www.vatican.va/content/francesco/en/encyclicals/documents/papa-francesco_20150524_enciclica-laudato-si.html

Race to Zero. (2021) GFANZ: Net Zero Financial Alliance Launches, Race to Zero. Available online at: https://racetozero.unfccc.int/net-zero-financial-alliance-launches/

Savio, R. (2018) Maurice Strong, Trail-Blazing Leader in Tributes and Reminiscences edited by Frederico Mayor, Negoslav Ostojic and Roberto Savio, European Centre for Peace and Development, University of Peace.

Schwab, K. (2018) Friends of Rio, Friends of Paris, Friends of Maurice Strong, in Tributes and Reminiscences edited by Frederico Mayor, Negoslav Ostojic and Roberto Savio, European Centre for Peace and Development, University of Peace.

Strong, M. (1992) Opening Statement to the Rio Summit (3 June 1992). Available online at: http://www.mauricestrong.net/index.php?option=com_content&view=article&id=165&Itemid=86

Strong, M. (2001) Where on Earth Are We Going? Texere.

Strong, M. (2004) Short Biography – University of Peace, Maurice Strong.net. Available online at: http://www.mauricestrong.net/index.php?option=com_content&view=article&id=15&Itemid=24

Strong, M. (2011) Statement by Maurice F. Strong at Special United Nations General Assembly Event on Rio+20, New York, October 25th, 2011, copy held by author of this chapter Felix Dodds.

United Nations. (1972) Report of the United Nations Conference on the Human Environment, UN. Available online at: https://undocs.org/en/A/CONF.48/14/Rev.1

United Nations. (1980) United Nations General Assembly Resolution 35/55 of December 5, 1980.

United Nations. (1987) Our Common Future: The World Commission on Environment and Development, Oxford University Press.

United Nations. (1989) UN Conference on Environment and Development: Resolution/Adopted by the General Assembly, A/RES/44/228, UN. Available online at: http://www.un-documents.net/a44r226.htm

United Nations. (1992a) Agenda 21 Chapter 28, UN. Available online at: http://www.un-documents.net/a21-28.htm

United Nations. (1992b) Agenda 21 Chapter 38, UN. Available online at: https://sustainabledevelopment.un.org/content/documents/Agenda21.pdf

United Nations. (1992c) Agenda 21 Chapter 19, UN. UN. Available online at: https://sustainabledevelopment.un.org/content/documents/Agenda21.pdf

University of Peace. (2018) Accreditation, University of Peace. Available online at: https://www.upeace.org/pages/accreditation

World Business Council on Sustainable Development. (2000) Changing Course: A Global Business Perspective on Development and the Environment, WBCSD. Available online at: https://www.wbcsd.org/Programs/People/Sustainable-Lifestyles/Resources/Changing-Course-A-global-business-perspective-on-development-and-the-environment

9

MARIA LUIZA VIOTTI

A Believer in Multilateralism and the Power of Collective Ideas: The Rio+20 Conference

Izabella Mônica Vieira Teixeira, André Aranha Corrêa do Lago and Luiz de Andrade Filho

Sustainable development multilateralism is often seen as a combination of successes and failures depending on how we look at it. Under the fleeting circumstances of the conclusion of a negotiation, governments rush to glorify accomplishments or justify frustrations. Civil society and other stakeholders celebrate achievements – discreetly at most – but will naturally point to their shortcomings. Meanwhile, the press works night and day to shed light on complex debates, and scientists talk through the gap between the best available knowledge and the modest outcomes.

Any objective view, nonetheless, has to recognize that it was precisely the action taken by the United Nations that has given the political, economic, and social relevance the sustainable development agenda has nowadays. Taking a long view allows us to see events and people from a different angle, where success is measured not in terms of missing commas or constructive ambiguities, but by how decisions have taken us closer to sustainable, healthy, and prosperous societies. Consensus, which for some is a source of frustration, is the vital strength of these negotiations, helping countries to move together – although often slowly – towards a better future on our shared planet.

This long view also provides an opportunity to celebrate the legacy of heroic individuals who have actively taken part in the construction of a comprehensive set of agreements ranging from climate change to protected species, chemicals management to desertification. But being a hero in multilateralism is something different from fiction: while fictional heroes pursue individual triumphs, heroes of multilateralism realize that collective success is about placing consensus above personal victories. This quality, as a matter of fact, seems to unite many visionaries that have over the past five decades come together to debate forward-looking ideas and to establish the building blocks of the UN sustainable development architecture.

DOI: 10.4324/9781003202745-10

Such real heroism is abundant in Ambassador Maria Luiza Ribeiro Viotti. Born in Belo Horizonte, Minas Gerais, Brazil, she bears some of the qualities that *mineiras* and *mineiros* are well-known for and are powerful consensus-building tools: the ability to listen and observe, and unwavering determination. A respected diplomat at Itamaraty (Brazil's Foreign Service), who made a brilliant career in multilateralism, she is the symbol of an institution that has continuously produced diplomats with a strong sense of duty and a worldview informed by complex (and sometimes conflicting) national interests – arguably the source of the insatiable desire of Brazilian diplomats to bridge positions in international negotiations.

Viotti has been a pioneer during her career: the only woman in the Itamaraty's 1975 batch, and the first and (up to now) only woman to serve as Permanent Representative of Brazil to the UN in New York, a position she held between 2007 and 2013. She was also the chairperson of the configuration of the Peacebuilding Commission for Guinea-Bissau – the first time a developing country presided over a mechanism of this kind. The time during which she led Brazil's representation to the UN coincided with international events of major relevance, including the deterioration of the security situation in multiple countries following the Arab Spring in early 2011 (precisely when Brazil was holding the Security Council presidency in February of that year) and the 2010 earthquake in Haiti, when Brazil was leading the UN Stabilization Mission in the Caribbean country (MINUSTAH).

When it comes to the environment and development agendas, the period Viotti led Brazil's Mission to the UN coincided with the proposal, negotiations, and immediate follow up of the 2012 UN Conference on Sustainable Development (Rio+20). She was instrumental in ensuring not only the convening of the Conference, despite some initial scepticism, but also that the long-nurtured concept of sustainable development would not get off track in the lead up to Rio+20. More importantly, that the immediate follow up of the Conference would be loyal to the spirit of the outcome document 'The future we want', particularly in relation to the negotiations for the elaboration of the Sustainable Development Goals (SDGs).

The adequate conceptualization of SDGs that followed Rio+20 represented a unique opportunity to elevate the sustainable development agenda, in its multiple dimensions, to an unprecedented degree of attention by the international community. To understand what was at stake for Brazil as the host of Rio+20 and the role played by Maria Luiza Viotti in the negotiations, it is worth going back to the origins of environmental multilateralism.

Once up on a time in Stockholm...

As strange as it may seem today, when the UN first started to respond to the growing concern over environmental threats, in the context of the 1972 United

Nations Conference on the Human Environment in Stockholm, there was a lot of scepticism around balancing an effective response to human-made impacts on nature with the fulfilment of societies' legitimate development aspirations. Environmental activists and development advocates were playing in opposite fields, and there seemed to be no chance of reconciliation.

Those contrasting worldviews between the North and the South clashed on multiple occasions and ways during the four-year preparatory process leading to the 1972 Stockholm Conference. But the hard work of delegates from 113 States, a capable presidency, and the active role played by Maurice Strong (see Chapter 8) and his team actually prevented the first act of this saga from having an abrupt end.

While some may consider the outcomes of Stockholm modest, its legacy informed processes and negotiations in response to local, regional, and global environmental challenges for years to come. The Vienna Convention for the Protection of the Ozone Layer, in 1985, its Montreal Protocol on Substances that Deplete the Ozone Layer in 1987, and the Basel Convention on the Control of Transboundary Movements of Hazardous Wastes and their Disposal of 1989, are noteworthy examples.

The Stockholm legacy also influenced the decision by the UN General Assembly to establish the World Commission on Environment and Development, composed of 23 representatives, including Maurice Strong himself. The Commission was tasked to develop what later became the *Our Common Future* report (also known as the Brundtland Report after its president, then prime-minister of Norway). Launched in 1987, the report placed environmental threats back at the centre of international debates. Most notably, it also coined the definition of sustainable development, which continues to be just as valid 35 years later:

> development that meets the needs of the present without compromising the ability of future generations to meet their own needs.
>
> *(UN, 1987)*

Once asked about the intellectual process behind the articulation of the concept of sustainable development, Indian economist Nitin Desai (a key expert in the Brundtland Report and later deputy-secretary general of the 1992 Earth Summit) explained that, from a conceptual viewpoint, the short statement had three crucial elements: the idea of needs, the ability to meet those needs and the link between the present and future capacity to satisfy needs. From a design perspective, Desai had a simple image in his mind: a bridge that had on one side environmentalists and on the other side development advocates. Both sides urgently needed a framework to be able to exchange ideas and better understand each other. Desai was convinced that only the UN had the legitimacy to build that bridge. Five years later, in Rio de Janeiro, it was to be that bridge's lead engineer.

Building the bridge: Rio, Part 1

The speech delivered by Brazil's President Fernando Collor de Mello at the opening ceremony of the 1992 Earth Summit caught the thousands of participants by surprise. According to Maurice Strong, in his book of memoirs, *Where on Earth are We Going?*, the President's words were:

> something of a surprise, so candid that he was about Brazil's environmental problems, including those affecting the Amazon.
>
> *(Strong, 2000)*

Collor de Mello called upon leaders to take bold action to raise the ambition in sustainable development, a concept "around which the rich and the poor, the big and the small can and must gather". At the same time, he clearly articulated the position of developing countries on key issues such as the need for new and additional financial resources. If there remained any doubts about Brazil's commitment to the success of the Conference, this carefully crafted speech probably dissipated them.

The outcomes of the 1992 Earth Summit are remarkable. In terms of numbers alone, it was the biggest event organized by the United Nations up to that point, bringing together delegations from 172 countries, 108 represented at the level of Heads of State or Government. The UN estimates the participation of 10,000 journalists and representatives from 1,400 NGOs. The Global Forum, a parallel event for civil society and other stakeholders, brought together over 7,000 participants. From getting the seating arrangements in the gala dinner for an unprecedented number of leaders to making sure those participants had their deserved roles in the process, everything was in some way new and experimental. Yet it ended up going extremely well, or at least without any major flaws.

The Conference's substantive legacy impresses even more than numbers – two legally binding instruments were opened for signature at the Summit: the United Nations Framework Convention on Climate Change and the Convention on Biological Diversity. Moreover, it launched negotiations on the Convention to Combat Desertification, open for signature in October 1994 (the three treaties are often referred to as the 'Rio Conventions', a source of joy for Brazilian diplomacy). The Conference also adopted three major international documents to guide future approaches to development: Agenda 21, an action-oriented programme containing strategies and recommendations to achieve sustainable development in the 21st century; the Rio Declaration on Environment and Development, a series of principles defining the rights and responsibilities of States; and the Statement of Forest Principles, a set of principles to underpin the sustainable management of forests worldwide. As a follow up to the Earth Summit, the Commission on Sustainable Development was created by the General Assembly later in that year to monitor the outcomes of Rio-92.

The tremendous success of the Earth Summit is attributed to a series of positive factors at the international and domestic levels. First, the momentum set by the end of the Cold War created favourable political conditions for multilateralism to prosper at levels only comparable to the immediate post-World War II. Environmental issues had by then gained much more attention with the press, civil society (seen on the exponential growth of NGOs and activists) and policy-makers, with concerns progressively shifting from localized environmental problems (such as transboundary water resources pollution, acid rain) to global ones (such as the depletion of the ozone layer and the correlation between rising CO_2 emissions and global warming), and therefore deserving of collective solutions.

Also at the international level, indigenous peoples and grassroots movements from developing countries became increasingly connected to international networks, particularly NGOs. Between the 1960s and 1980s, environmental concerns resonated mostly with a few sectors of civil society in the wealthiest countries. From the 1980s on, demands from grassroots movements became progressively linked to the international environmental agenda, thus making the case for greater action by governments in both developed and developing countries.

This was the case of Brazil, which at that time was witnessing events that would forever change the political relevance of environmental issues: the Amazon rubber tapper's demands for land rights led by activist Chico Mendes (brutally murdered in 1988), and the debate on embedding indigenous rights and the protection of the environment as Constitutional rights in the context of the democratic transition – which eventually happened in 1988. Brazil in many ways epitomized the challenge and the opportunity that the Earth Summit vision represented: a country facing many challenges, from poverty to economic recovery, yet home to the Amazon and the world's largest biodiversity.

This background very much influenced Brazil's approach to the Earth Summit, including the bold decision to host the Conference, becoming an active voice in the sustainable development regime, and to establish an inclusive preparatory process at the national level that brought several governmental bodies together and gave a special role to representatives of businesses, academia, and NGOs as observers.

The legacy of the Earth Summit shaped a generation of multilateralism champions in Brazil's foreign policy, including Maria Luiza Viotti. Those individuals had witnessed the profound changes in Brazilian society from the 1980s and how sustainable development, including the fight for indigenous people's rights and nature conservation, became part of the country's democratic transition. The generation of 1992 became the leadership in 2012, and that profoundly influenced how Brazil approached Rio+20.

Crossing the bridge: Rio, Part 2

"What is happening in Brazil? A revolution?" So reacted Germany's former Chancellor Angela Merkel while walking past the table of the Brazilian delegation at the 2010 Millennium Development Goals Review Conference. Seated behind the flag were five Brazilian women leading on the debate of a vision for development at the end of the first decade of the 21st century. Ambassador Maria Luiza Ribeiro Viotti was the chief negotiator (Maia et al., 2010).

Brazil was not going through a revolution as such, but society and the economy were much different from what they had been in 1992. The country had reduced the population living in poverty from 20% to 5%. While diversifying its economy and improving agriculture productivity, Brazil was making substantive progress on the environmental agenda, working decisively to bring down deforestation rates in the Amazon. The country was one of the very few to generate more than two-thirds of its power from renewable sources. Sustainable development seemed closer to reality.

Those major changes help to explain the surprising proposal not only to convene but also to host another international conference on the environment and development. What came to be Rio+20 was first proposed in 2007 in the Brazilian Parliament by former President Collor de Mello – then a Senator –to commemorate the legacy of the Earth Summit. The idea was well accepted in the Brazilian Government, quickly gained traction, and was eventually introduced in the UN. For many analysts, the very fact that it had been a developing country to launch the idea of such a conference was proof to the political relevance and importance the sustainable development agenda had acquired globally.

What informed Brazil's idea of convening and hosting Rio+20? Some viewed the country trying to strengthen its "emerging" status, carrying out a policy strategy to be a bridge between the developing and the developed worlds. To others, Brazil was seeking to reaffirm its credentials as a balanced leader determined to strengthen multilateralism. In reality, however, the main political drive to propose Rio+20 was the realization that Brazil's 1972 long-standing assessment of the need for a balanced and integrated approach to the economic, social, and environmental dimensions of development, was more valid than ever. An event of the magnitude of the Earth Summit could help prevent multilateralism from backsliding on the environment–development nexus, while at the same time maintain the active engagement of developing countries, including Brazil's, in this agenda.

However, there was some scepticism around what such a Conference could achieve. International circumstances were very different from the early 1990s. While the Earth Summit in 1992 took place at the apex of multilateralism, the proposed sustainable development conference would happen in a world impacted by the severe negative effects of the 2008 financial crisis – one originated in developed economies that led to political, economic, and social instability. Multilateralism was under severe scrutiny, still suffering from the consequences of how the UN had handled the situation in Iraq in the early 2000s (Corrêa do Lago, 2009).

Overcoming this hurdle was Maria Luiza Viotti's first task. She had an active role in getting the mandate for the Rio+20 Conference at the UN General Assembly in New York, which eventually happened in December 2009. Starting in late 2008 and throughout the following year, she conducted a series of small, informal meetings with other permanent representatives and civil society stakeholders to explain Brazil's rationale for convening a new conference on sustainable development. The Republic of Korea, which had initially showed interest in hosting the conference, eventually decided to drop its candidature and became a key ally of Brazil in promoting Rio+20.

During this process, Viotti articulated a vision to create momentum for the convening of the conference in those challenging times. First, the then ongoing financial crisis was a reminder of the need to rethink the role of the State in shaping growth. Second, the United States' renewed interest in multilateralism represented a unique opportunity to achieve consensus. Finally, the Conference should explore a forward-looking vision for sustainable development that could bring developing and developed countries closer together.

Against this backdrop, while the Earth Summit had been planned to be a "point of arrival", with the conclusion of a series of multilateral agreements and declarations, the Rio+20 Conference was conceived as a "point of departure", proactively launching a series of new processes and negotiations.

This forward-looking mandate was appealing, but some saw it with reservations. Multiple countries raised questions, for example, about the mobilization of Heads of State and Government to a Conference that could potentially be scarce in results. Viotti carefully dealt with those concerns, encouraging countries to own the idea of the Conference, to collectively refine its concept and establish priorities and, more importantly, to work decisively towards its success.

Brazil's optimism and Viotti's determination paid off. Rio+20 is considered one of the greatest achievements in multilateralism of the past decade, having delivered both on launching important new mandates and on the reform of UN environment and development governance structures. In terms of new negotiations, Rio+20 initiated a process that came to successful conclusions in the following years: the development of a new set of Sustainable Development Goals in 2015, building and expanding on the model of the MDGs; and an intergovernmental process for consideration of future development finance needs – concluded in Addis Ababa, also in 2015, during the 3rd International Conference on Financing for Development. Combined, both became Agenda 2030 – the most comprehensive and ambitious development agenda ever established, adopted at the level of Heads of State and Government in New York in September 2015.

As for governance, Rio+20 mandated the establishment of a High-Level Political Forum on Sustainable Development, of universal participation, which succeeded the Commission on Sustainable Development that met annually since 1993. The HLPF is now the main UN platform on sustainable development and has a central role in the follow up and review of the 2030 Agenda, including through regular voluntary national reports. Rio+20 also decided to strengthen and update UNEP, with the establishment of its Environmental Assembly – UNEA, now the highest-level decision-making body on the environment, with universal participation and biennial meetings, succeeding the former UNEP's Governing Council which was composed of less than 60 Members.

Other important outcomes include the development of guidelines for the advancement of the green economy in the context of sustainable development, and the adoption of the 10-year framework of programmes on sustainable consumption and production patterns.

Finally, Rio+20 was also successful in integrating civil society into the multilateral process, through the innovative 'Dialogues for Sustainable Development'. Conceived by the Brazilian government, the Dialogues engaged more than 60,000 people in virtual and face-to-face discussions on priority themes of the international sustainable development agenda, through an online debate platform created in partnership with the UN Development Programme and coordinated by 30 Brazilian and foreign universities. By public vote and by nomination of the debaters, 30 recommendations (three per panel) were chosen and taken to Leaders.

This pioneering multistakeholder format, unprecedented in UN conferences, contributed to a qualitative improvement of civil society's participation in multilateral processes on sustainable development.

But getting to those comprehensive results was no easy task. During the early negotiation stages, many analysts were quick to extrapolate the failures of the 2009 Climate Change Conference in Copenhagen (see Chapter 6) to environmental multilateralism in general. Echoing this view, some developed countries advocated a thorough review of the principles contained in the 1992 Rio Declaration, including the principle of common but differentiated responsibilities – a non-starter for developing countries.

Viotti skilfully led the tough negotiations to establish the mandate for Rio+20 and its thematic issues, while preventing the Conference from re-opening the Rio Declaration. She led an incredible and diverse team of diplomats that worked tirelessly to provide headquarters in Brasília with timely and balanced information about debates in New York, combined with sharp proposals on how to overcome difficulties.

Civil society stakeholders were key partners in the Rio+20 journey from the outset. The ever-more connected networks of NGOs, activists, and associations embraced the idea of a new conference on sustainable development and wanted to actively contribute to its success. This fruitful interaction between networks – often facilitated by the Brazilian Government – resulted in creative ideas on how major groups could be an active force in the Rio+20 process, including through the Dialogues for Sustainable Development.

The process for the creation of the SDGs got special emphasis, given its potential to be one of its key deliveries. In April 2011, during one of the informal meetings held in Rio, Brazil decided to introduce on the negotiating agenda the idea of a new set of development goals in tune with nature. The idea had been first presented in the High-Level Panel on Global Sustainability, a group of personalities invited by the UN Secretary General that worked in parallel with the

Conference.[1] Brazil's objective was to incorporate discussions around the SDGs into the preparatory process, so that these were reflected in national submissions to be presented in November 2011, therefore giving legitimacy to its inclusion, by the Secretariat, in the initial draft negotiating text. The success around the decision on the SDGs mandate in Rio de Janeiro is due, to a large extent, to the collaboration maintained with Colombia throughout the entire process.

Here, Viotti also made a significant contribution. Having closely followed the UN work on supervising the implementation of the MDGs, she ensured that the mandate to establish the SDGs were designed to substitute the MDGs and go beyond them. This included building a coalition of like-minded countries in New York that successfully promoted the idea of universal goals, applicable to both developing and developed countries. The underlying vision of universality represented a major change, acknowledging that all countries faced challenges to fulfil development needs. A common development programme of work would make that clear. The successful negotiations that led to the adoption of Agenda 2030 in just three years proved that a decent dose of pragmatism and idealism are usually a good formula for effective multilateralism.

Brazil recognized early on that the Rio+20 formal preparatory process was significantly shorter in comparison to the 1992 Earth Summit. That is why a series of inter-sessional and informal meetings were convened to make progress in between the formal meetings, focusing on the key issues which the Conference had to deliver on – mainly the SDGs mandate and environmental governance, including the future role of UNEP. Leading on the day-to-day preparatory process of Rio+20 in New York, including by working closely with the bureau in Brazil's host capacity, Viotti worked to provide a clear picture of what was realistically achievable, without losing sight of new ideas or creative ways of unblocking consensus. This dynamic created a virtuous cycle in key moments of the negotiation, with informal meetings pushing the boundaries of ambition in New York.

Despite many formal and informal negotiation sessions, only about one-third of the negotiating text was cleared during the last formal meeting, already in Rio de Janeiro (13 June 2012). Brazil conducted informal negotiations during the days that preceded the High-Level Segment. On the evening of 15 June, the presidency started to work on it, making the necessary political choices based on discussions. After intense consultations and drafting and revision work, on the morning of 19 June, the Brazilian presidency tabled a proposal carefully developed by a group of senior Ambassadors, including Viotti, and their aides who were following discussions in key negotiation rooms. The proposal was endorsed at chief-negotiators level at around 1:00 pm on that same day and eventually adopted by Heads of State and Government on 22 June.

The conclusion of negotiations before Heads of State and Government arrived in Rio, through an open, transparent, and inclusive process, reiterated the commitment of the Brazilian Presidency to the success of negotiations overcoming differences through balanced compromises – just like in 1992. Brazil was determined to demonstrate that international conferences of that magnitude can and must result in text agreed at the negotiators' level, without imposing on leaders the burden of directly negotiating texts – an unrealistic task, as the Copenhagen experience in 2009 had clearly demonstrated. For Brazil, this collective achievement – the adoption of a comprehensive, conceptual, and operational document by consensus in a political atmosphere much different from 1992 – represented the revitalization of multilateral processes.

The Rio+20 outcome document, 'The Future We Want', was the conceptual and political starting point to establish an agenda for sustainable development for the 21st century. It effectively crosses the bridge that the Brundtland Report experts envisioned: the three issues mentioned as the main priorities agreed by the international community are, first, the eradication of poverty; second, the change in unsustainable patterns and promotion of sustainable consumption and production patterns; and, third, the protection and management of natural resources that are the basis for economic and social development. Those three priorities translate, in an exceptional way, the need for integration of the three pillars of sustainable development, but they also individually show the preponderance of one of the pillars over the others. In Rio in 2012, the international community unequivocally agreed to a development agenda that was realistic and ambitious, but at the same time visionary and idealist.

The future we want after Rio+20

The process to establish the SDGs is largely seen as the main legacy of Rio+20, in line with the "point of departure" approach that the Brazilian Presidency had envisioned. The possibility of combining the goal-oriented approach inherited by the MDGs with a strong element of universality appealed to both developed and developing countries. In many ways, the programmatic dimension of Agenda 21 coupled with metrics had the potential of putting the sustainable development agenda at the heart of the international agenda, particularly in an increasingly climate-centric approach to development challenges.

But the mandate established at Rio+20 for an inter-governmental process to define the SDGs inherited a mismatch that would lead to very hard negotiations from the outset. The consensus arrived at in Rio+20 instructed the establishment of "an inclusive and transparent intergovernmental process on sustainable development goals that is open to all stakeholders" with an open group comprised of 30 representatives. The open group was further given the task to develop methods of work that "ensure the full involvement of relevant stakeholders and expertise from civil society, the scientific community and the United Nations system in its work, in order to provide a diversity of perspectives and experience".

This compromise reflects diverging views on the envisioned process to arrive at a new set of development goals to build on the MDGs. For developing countries, while there was wide recognition that the MDGs had contributed to advancing the development agenda at the international level, there was also a strong criticism of how they had been shaped – behind closed doors, predominantly by experts of developed countries, which ultimately would not be responsible for their implementation. That's why, according to that view, the SDGs should follow a different path, being developed in a widely participatory manner.

For developed countries, on the other hand, there was a perception that a large negotiating group would enhance the North–South divide, with a risk of focusing excessively on financial resources. In the aftermath of Rio+20, some countries even floated the idea of a two-track process, with an MDG+ set of goals exclusively for developing countries and universal SDGs encompassing all countries.

Settling this "conflicting mandate" of being fully inclusive while also led by just 30 countries was no easy task. It was not in anyone's interest – not least Brazil – to risk the legacy of Rio+20 and the unique opportunity that the elaboration of the SDGs represented.

It was Maria Luiza Viotti who facilitated the extremely hard negotiations that eventually led to the adoption of the work methods of the negotiation group and its composition. She adopted the posture that such complex negotiations require, creating conditions for Member States to realize by themselves – rather than through imposition – that only a creative solution could fix the paradox that the Rio+20 mandate had created.

Building on the experiment of the Green Climate Fund, at whose governing board countries were working through constituencies, she introduced the idea that those 30 seats could be shared during the forthcoming negotiations in order to ensure inclusive participation. In addition, the working group was eventually made open to any interested Member State, which could take the floor and contribute substantively to discussions.

The final composition of the group is a must-see for multilateralism enthusiasts, with seats shared by two, three, and sometimes even four countries. Some of them often did not share similar views or approaches to international issues but had through the open working group the opportunity to work more closely together. Not only at this point but throughout the long and complex Rio+20 process (from its original idea to negotiations and follow up in New York), Viotti used her diplomatic skills to ensure the process would not fall into polarizations and divisive passions or emotions running high.

Maria, Maria

As many diplomats know, diplomacy is a more powerful tool when employed to avoid crises rather than to solve them. An experienced Brazilian Ambassador was once asked by his children about the nature of the job and why it was different from pure politics. After thinking for a few minutes, he explained that to be a good diplomat, someone must have a personality that can live without public recognition. He elaborated on that idea: the main objective of diplomacy is to achieve and maintain peace, so the essential job of a diplomat is (1) to identify a situation that can lead to a crisis before it becomes one; (2) try to diffuse it; (3) prevent the situation from becoming public (the most difficult part), and finally; (4) to make sure that very few people know how you avoided it. The more you talk about the role you played in avoiding a crisis, the sooner will people think it is a non-issue (since it did not happen!) or – even worse – will bring it back.

Politicians, on the other hand, tend to tackle crises when they appear (sometimes with the help of diplomacy), but diplomats avoid innumerable crises that might have given lots of headaches to politicians (and even wars) – but no one knows – he concluded.

Maria Luiza Viotti is considered by her peers a paradigmatic example of that classic and effective diplomat. She is regarded by many of her colleagues as a highly skilled, knowledgeable, and self-assured negotiator – a discreet and powerful style of leadership. Many of her outstanding achievements in multilateralism have not been widely reported simply because she has an extraordinary talent for identifying challenges much ahead of the crisis curve. And she will most likely succeed in avoiding them. Countless times.

Those are skills she continues to demonstrate as Chef de Cabinet of UN Secretary-General Antonio Guterres. In her current position, she has gained wide respect amongst UN Member States for being a reliable aide of the Secretary-General during the past few years, including for advancing his ambitious proposals to reform the development, management, and peace and security pillars of the UN system. As a firm believer in multilateralism and a proud member of the UN family, Viotti has been a key ally in reassuring the vital role of the organization for generations to come. She is one of those family members that provides stability. Someone to talk to when looking for good advice to avoid crises or ideas to get out of crossroads.

As multilateralism continues to demonstrate its importance, our common future cannot but be seen through the lenses of sustainable development: from the food we eat to the waste we generate, from building bridges and schools to designing financial products and technology solutions, from fighting corruption to promoting indigenous rights, environmental, social, and economic concerns have become inextricably linked to one another. The level of importance of sustainable development in international relations nowadays is largely the result of the manner in which the subject was tackled at the multilateral level.

The 50-year-old journey of environment and development in the UN may seem long, but it is far from concluded. Our heroes, like Viotti, will continue to inspire present and future generations to make sure that we stand the chance of living healthy, safe, and prosperous lives in tune with nature. Sustainable development multilateralism in the crucial decades to come may well be represented in 'Maria, Maria', a masterpiece composed by Milton Nascimento, one of Brazil's most celebrated artists born in Rio de Janeiro and raised in Minas Gerais. Maria represents "people that smile, even when they should cry" and that "always have dreams". Maria is "a force that warns us, but has faith in life".

Note

1 Two of the authors, Ms Teixeira and Mr Corrêa do Lago, were respectively Member and Sherpa of the High-Level Panel on Global Sustainability.

References

Collor de Mello, Fernando. (1992). Discurso pronunciado pelo Presidente Fernando Collor por ocasião da abertura da Conferencia das Nações Unidas sobre Meio Ambiente e Desenvolvimento, no Rio de Janeiro, em 3 de junho de 1992 in Resenha da Política Exterior do Brasil. Ministério das Relações Exteriores. http://www.funag.gov.br/chdd/images/Resenhas/RPEB_70_jan_jun_1992.pdf

Corrêa do Lago, André. (2009). *Stockholm, Rio, Johannesburg: Brazil and the Three United Nations Conferences on the Environment*. FUNAG.

Corrêa do Lago, André. (2013). *Conferências de Desenvolvimento Sustentável*. FUNAG.

Maia, Melina E. and Alvarenga, Tainá G. (2010). *Marias do Brazil: as mulheres e a diplomacia pelo olhar das Embaixadoras do Brasil na ONU* in JUCA: Diplomacia e Humanidades. ISSN 1984–6800. http://www.institutoriobranco.itamaraty.gov.br/images/pdf/Juca/04/JUCA-04-INTERNET.pdf

Strong, Maurice. (2000). *Where on Earth Are We Going?* Texere Press.

United Nations. (1987). *Our Common Future*. The World Commission on Environment and Development, Oxford University Press.

United Nations. (2012). *The Future We Want*. Outcome Document of the United Nations Conference on Sustainable Development. https://sustainabledevelopment.un.org/content/documents/733FutureWeWant.pdf

10

PAULA CABALLERO

Building a Blueprint for a Better Future: The Sustainable Development Goals

Irena Zubcevic

A sigh of relief was heard from all corners of Conference room 2 in the basement of the United Nations Headquarters in New York one day in December 2009. The Chairperson of the Second Committee of the United Nations General Assembly, Permanent Representative of South Korea In-kook Park, had just gavelled through a resolution on the implementation of Agenda 21.[1] The resolution, which in most years would contain recommendations for the UN Commission on Sustainable Development (CSD),[2] this time contained something quite different. It was an important decision that had been in the making since 2007 (UN, 2009).

Just what was this decision, exactly? Why did it matter? And how did a Colombian government official, Paula Caballero, come to play such an important role in what happened next?

How it all began

The decision was to hold the United Nations Conference on Sustainable Development (UNCSD)— better known as 'Rio+20'—in Brazil in 2012.[3] Developing countries, spearheaded by Brazil[4] and South Africa,[5] which had previously hosted major conferences and summits on sustainable development, were happy with the outcome and the selection as one of the two themes: an institutional framework for sustainable development. The other theme, on a green economy in the context of sustainable development and poverty eradication, was a compromise to get the European Union and other developed countries to agree to the Conference. The concept remains controversial to this day however, with many developing countries still saying there is no common understanding of the term and fearing that official development assistance and market access could be made conditional on adherence to this ill-defined concept.

DOI: 10.4324/9781003202745-11

The decision took a long time to finalize. It started to be discussed during the 62nd session of the General Assembly in 2007 in the context of the negotiations on the Agenda 21 resolution after the then President of Brazil Luiz Inácio Lula da Silva proposed 'a new Conference on Environment and Development to be hosted by Brazil in 2012' in his speech at the UN General Assembly general debate.[6] However, the decision did not make it into the final draft that year or the following year. Thus, countries of the G-77—the UN name for developing countries in the world[7]—and Brazil in particular, started lobbying to have a summit on sustainable development 20 years after the first Rio Conference in 1992. However, developed countries, in particular the European Union (EU) and the USA, opposed what they called 'yet another summit' especially as they felt CSD had a multi-year work programme running until 2017 and it would make more sense to have a conference then. In December 2008, in resolution 63/212 (UN, 2008), the EU and US succeeded in deferring the decision for yet another year, inviting Member States to express their views on the possibility of convening what they called at that time 'a high-level event on sustainable development' and requested the Secretary-General to report on the views in his report on Agenda 21 to the next session in 2009.

It was a difficult time for the world. The 2008 financial crisis had just occurred, and developed countries were not keen on taking on new commitments. However, the majority of views received by the Secretary-General supported the idea of the Conference and in the hindsight, it could be seen that holding a conference in 2012, and not as the US proposed in 2017 at the end of the multi-year work programme of the CSD, was the right decision. It was fortunate that Brazil had succeeded in convincing delegations that CSD was no longer delivering what it had promised and that a conference could look at emerging issues and new solutions for sustainable development. Had the CSD limped along for another five years, the SDGs would not have materialized. It is anyone's guess what if anything would have taken over from the Millennium Development Goals (MDGs) whose achievement date was 2015.

And so, the delegates that December 2009 in New York were finally able to go on their holiday break and leave negotiations behind. They could be forgiven for thinking that the content and outcome of this conference in 2012 was several years off and might be taken up by a different set of delegates.

Planting SDG seeds

However, almost 2,500 miles south of New York, in a beautiful 16th-century San Carlos palace in Bogotá, there was one person who understood that having a conference was not enough. As part of the preparatory process for Rio+20, in the fall of 2010 the UN sent a survey to all governments asking them what expectations they had for this conference. Colombia, like all other countries, received the survey, and the task of completing it fell to the newly appointed (in October 2010) Director of Economic, Social and Environmental Affairs at the Ministry of Foreign Affairs of Colombia, Paula Caballero. She was not impressed by the resolution (UN, 2008) adopted in New York. She found the themes vague, especially the one on green economy. She knew there were conflicting views as to what it meant. This view was shared by Farrukh Khan of Pakistan[8] who later became a bureau member for Rio+20. He said that there was a feeling among many G-77 countries that the themes for Rio+20 were very much driven by the OECD (western) countries and the UN system, especially the green economy, sustainable production and consumption patterns and sustainable energy for all, and that they regarded Rio+20 as a conference to address these issues.

Paula Caballero did not want to have yet another politically negotiated document with little implementational value, as she thought had been the case with the outcome of the World Summit for Sustainable Development in 2002. She felt that this could be a missed opportunity to use a high-level summit to generate traction and pathways for substantive change.[9]

Caballero came from a family of thinkers and inventors. Her great grandfather built a telescope to watch sunspots. And she herself worked on many projects related to watersheds and fisheries with concrete outcomes that benefited people on the ground. This was a knowledge and experience that would serve her well on her path to conceptualizing the SDGs.

With the support of the then Colombian Vice-Minister Patti Londoño Jaramillo, Caballero gathered in January 2011 a group of colleagues in a freezing cold room in the Ministry overlooking a patio with palm and oak trees, where she shared her views and concerns. 'We need something short, pithy and compelling', she said. 'We need something like the MDGs[10] as they galvanized everyone, but we need it for the whole of development to embrace its complexity', she elaborated. And so, after a few hours of discussions, the idea crystallized that goals for sustainable development were what was needed.

Caballero was influenced by her own childhood growing up on the outskirts of Bogotá in the mountains. It has shown her first-hand how people and nature are connected and how people need to live in harmony with nature to achieve better lives. This has served her well in her thinking when she planted what would become the first seed which eventually grew into the SDGs. She showed this very first proposal to her minister, Patti Londoño Jaramillo. The minister liked this concrete proposal and told Caballero, 'We need to make it happen' and pinned it on her otherwise bare wall. And she did, providing ministerial support to Caballero throughout the process together with the then foreign minister María Ángela Holguín. This political support was invaluable as their work on the SDGs progressed. It showed that it was not only Caballero's own idea but had the full support of her government.

So while the wind was blowing around the cold streets of New York in February 2011, Caballero embarked on a long and arduous journey towards achieving her own dream to improve the lives of all people. She wanted to encourage people to stop working in silos and to force everyone to see development as a comprehensive whole that had to be managed as such. She was increasingly convinced that a more comprehensive metric could be vital in triggering action and getting more integrated responses to the many development challenges we were facing.

Conceptualization of the SDGs

Caballero started her first effort to inform others of her idea using the ninth session of the United Nations Forum on Forests that took place in February 2011 to get reacquainted with negotiators in the sustainable development arena. In the corridors of the United Nations Headquarters in New York she started talking to any delegate, UN staff or stakeholder who wanted to listen. She also thought it would be much easier if she could get a few countries from the Latin American region or the whole region understand her ideas around SDGs. However, at that time only two delegates in their own personal capacities, Jimena Leiva of Guatemala and Ye-Min Wu of Singapore worked with her closely. They suggested to link the SDGs more strongly with Agenda 21 to demonstrate that they were fully embedded in Rio 1992's legacy.

It was too early even for those who saw the value in the SDGs to fully embrace the idea that SDGs would in fact replace MDGs. But not Caballero. Having worked in UNDP and on many projects on the ground, she understood the value of the MDGs in providing a valuable framework for countries' development. However, she also understood that the MDGs were not enough to truly secure sustainable development globally and that the development and challenges addressed in the MDGs were too limited in focusing on developing countries.

Turning point

This was a turning point. Caballero went to meet colleagues at The Pod 51 Hotel, a modern hotel close to United Nations Headquarters. Caballero wrote a new version of the SDG paper at the Pod based on her discussions, saying:

> Colombia is proposing that a key outcome of the 'Rio +20' process be the definition and agreement of a suite of Sustainable Development Goals (SDGs), equivalent to the MDGs. These SDGs would translate the Green Economy/Sustainable Development debate into tangible goals, which would focus the broad debate at a practical level and enable the preparatory process to productively address key issues for which measurable progress would be welcome.
>
> *(Colombia, 2011)*

This helped her discussions and became the basis for advocacy over the next seven months. It added more specific areas that the SDGs could address and also included an important proposal on the process post-Rio+20 by which the SDGs would be defined and agreed. And it was based on an historic merging of the two parallel trends in development: one embedded in structural economics and the other one in sustainable development.

These two streams had existed in parallel for decades. The one around sustainable development was very much marginalized and relegated to sustainable development conferences, starting with Stockholm in 1972[11] and the CSD, which especially towards its end was considered the realm of exclusively environment ministers and environment stakeholders. Meanwhile, what was considered 'core' development was focused around the MDGs, both in countries and in international organizations like the World Bank and IMF through its poverty reduction strategy papers (PRSPs).[12] The United Nations had its UN development system headed at that time by the UN Development Programme (UNDP) and its environment-focused system headed by UN Environment Programme (UNEP).

Within the UN Secretariat, the Department of Economic and Social Affairs (DESA) was the secretariat for Rio+20. Within DESA, the leading economists— Jomo Kwame Sundaram, Assistant Secretary General for Economic Development and Robert Voss, Director of the Development Policy and Analysis Division— had led analytical work embedded in structural economics. However, they started recognizing in 2010 that eventually development progress would slow, and in some cases, even be reversed unless greater attention was given to global challenges like climate change and thus, increasingly featured sustainable development dimensions. As coordinators with UNDP of the UN system team which produced the UN report on the post-development agenda, they succeeded in broadening the scope of the report well beyond that of the MDGs (see below), which likely was a key factor in later securing support to the idea of a convergence between the post-2015 and sustainable development agendas within the UN system.

Within DESA though, the main support for the concept of sustainable development and the legacy of Agenda 21 came from the Division for Sustainable Development (DSD), which had supported the Commission on Sustainable Development since its inception. DSD was the main division in DESA supporting the preparations for Rio+20. A multi-disciplinary team led by David O'Connor[13] was writing the official documentation (Secretary-General's reports) for Rio+20 and prepared a comprehensive assessment of progress since the Earth Summit. The team also produced a number of policy briefs prior to Rio+20 (UN, 2012a), including one elaborating on SDGs, where the proposal of Colombia and Guatemala was discussed and where the idea of measuring the SDGs was introduced. In part because it was directly supporting the negotiations on the outcome document of the Conference, the Division could not engage in advocacy as other parts of the UN system were doing, and its work received much less attention both within and outside the UN system.

In the UN system, from 2011 on, there started to appear discussions on the 'post-2015 United Nations development agenda', 2015 being the year when the goals set by the MDGs were supposed to be met. This was reflected in the paper in May 2011 produced by Paul Ladd who worked at UNDP (Dodds et al., 2017) and later in the UN Secretary-General's annual report from July 2011 on accelerating progress towards MDGs (UN, 2011). Those were the first attempts for at least recognizing that something more than the MDGs was needed.

Another project under the radar that could have helped inform the discussions prior to Rio+20 was the one that resulted in the report on 'Sustainable Development in the 21st century' which was composed of a series of studies on assessment of progress since the Earth Summit; emerging issues; long-term sustainable development scenarios; tools for managing sustainable economies; national and international institutions for sustainable development; and sector assessments. This study, funded by the European Commission and conducted by DSD was the first major study that started discussing interlinkages and trade-offs and is still relevant today. It strongly underlined that

> There is no single solution or policy for sustainable development. Bottom-up measures and policies need to be tailored to each issue, country, and sector. Great differences remain in terms of specific policy recommendations that are drawn ex-post from scenario results. A key problem is the existence of important trade-offs across time, sectors, and issues. Scenarios produced for Rio+20 also highlight the equally important synergies and opportunities provided by policy strategies that are geared to simultaneous achievement of multiple sustainable development goals.
>
> *(UN, 2012b)*

Both countries and stakeholders followed the same streams. Among stakeholders, there were those who were traditionally supporting sustainable development through CSD, called Major Groups, that were established in Rio 1992[14] and the developmental stakeholders supporting MDGs and financing for development processes.

Caballero said that questions were continuously asked about why she should be advocating for a new set of goals when the MDGs were still 'unfinished business' and unlikely to be fully met by 2015. After all, it was still early 2011, they said. Many officials both in countries and within the UN system thought it would be much better to roll over the MDGs after 2015, with a few minor adjustments. This was what came to be known as MDGs+.

This is not surprising. As human beings, we often tend to live in our comfort zones. Development programming systems at both national and international level had come to be defined around measuring progress towards the MDGs. They offered a comfort zone for many developing and developed country governments, the UN system and other international and regional organizations. As Khan said,[15] it took at least half of the 15-year timeframe of the MDGs to figure out how to place them at the centre of development policy and cooperation efforts. And so around 2007, when everyone in the development community had figured it out, they felt comfortable with the MDGs. Structures were in place nationally and globally to work on MDGs and it was only natural to stick to something familiar. It was also true that many countries were making too slow progress on poverty and hunger eradication and various health and education goals, so they felt uncomfortable faced with something new when there was still unfinished business with MDGs.

However, countries also understood that something had to come out of Rio+20 process and sticking to the known framework and adding a little bit more, mostly around agriculture, energy and water was not something Caballero was supportive of, because she considered that there was a need for a more explicit focus on sustainability. She was also aware if the SDGs would not be accepted in Rio, MDGs that had a separate track even in the MDG+ version, would not have been part of the 'Rio+20 outcome'. But with the SDGs, things were about to change. However, there was still a long way to go as many, especially developing countries and developmental NGOs, felt that SDGs through its universal character might divert attention and resources from their core focus on poverty eradication and economic growth.

And so, Caballero's journey continued. She saw her country and herself as bridgebuilders and for her the SDGs were 'an offering we put forward to create a single common agenda that entailed collective responsibility and at the same time, collective empowerment. We wanted to catalyse a sense of shared destiny'. She spent numerous hours talking to delegations, UN system representatives and stakeholders in what was known as the North Lawn Building at the UN in New York, a soulless, temporary construction (thankfully dismantled since) used for meetings while the UN building was being remodelled.

The building did not look conducive to any kind of cozy talks, as it was built of cement and the floors were concrete slabs. In short, being inside it felt like standing on a street or sidewalk. But there was at least the Vienna café on the second floor. It was not as cozy as the original Vienna café, housed in the permanent Headquarters' basement. That original café was an excellent place for meeting and networking and some people were almost fixtures there. For instance, Felix Dodds[16] could be found ensconced in the Vienna café, often chatting with delegations or strategizing, while delegations were found resting on the couches when negotiations went on until the wee hours of morning. This new Vienna café in the temporary building was much less inviting, but delegates and stakeholders still gathered there. There was nowhere else to go. Taking advantage of this fact, Caballero started spending time there, talking about the SDGs with anyone who would listen.

First milestone

The first milestone happened on 27 May 2011 when the very first informal intergovernmental consultation on the SDGs took place. It was held at the Permanent Mission of Colombia on 57th street of New York's Upper East Side. About 20 people came representing all UN regional groups.[17] Caballero recalls that her proposal was met with friendly scepticism but no clear support. A few expressed mild interest but considered it would be impossible to get agreement around it. With the support from her Government and the Permanent Mission in New York that facilitated negotiations, organized events and maintained high levels of advocacy at the UN in New York, she used every possible occasion to promote the proposal. In June 2011 on the margins of the UN's climate change negotiations, she spoke to the Brazilian head of delegation, Andre Correa do Lago. He expressed his interest because he knew that as a host Brazil has a unique role to play. And the SDG proposal was definitely discussed in the corridors and more colleagues came to express support. Still, the situation was fluid and there was no unanimity of views even within the same government or a UN entity. Nevertheless, Caballero felt that the time was ripe to officially table the proposal. It was only a year before Rio+20.

SDGs officially on the table

During the High-Level Dialogue on the Institutional Framework for Sustainable Development held during 19–21 July 2011 in Solo, Indonesia, Caballero officially presented a proposal by Colombia on the SDGs. This was the first time that the SDG option was formally proposed at a UN event and mentioned in the Chair's summary from the meeting, even though it was not officially discussed at the meeting itself. But it was discussed informally in corridors and cafes. 'Those were rich and substantive discussions. I recall talking to the whole EU delegations in an informal way', recalls Caballero. Stakeholders, and in particular Felix Dodds,[17] then Director of the Stakeholder Forum for a Sustainable Future, played an important role in Solo supporting the SDGs.

Another milestone came during the first of the two informal consultations hosted by Brazil in the run up to Rio+20. This consultation was held in Rio de Janeiro's neoclassic, almost intimidating Palácio Itamaraty, which until 1970 was the headquarters of the Brazilian Ministry of Foreign Affairs. The consultations started on 21 August 2011. By that time, it was quite evident that there was no excitement about Rio+20 and that no one was expecting much.

That the SDGs provided something new and a real buzz of excitement could be discerned when Caballero started talking about them. Discussions were fruitful. They prompted her to rewrite the proposal from the day before so that it was not linked so much to Agenda 21 but focused solely on the concept of SDGs. She shared a new version of the paper on 22 August after little sleep as the delegations were gathering for the second day of discussions. This proposal was also endorsed by Guatemala. According to Caballero, Brazil did not explicitly endorse it, but Ambassador Luiz Alberto Figueiredo of Brazil, who headed the negotiations on the outcome of the Rio+20 from Brazilian side, indicated towards the end of the meeting that 'Rio+20 could not focus only on principles and that concrete deliverables were also needed'. Caballero interpreted this as a tacit support for a very tangible proposal. Even though a couple more versions were to follow, it is this paper that most remember when recalling the process at the time as the basis for the proposal that was finally included in the outcome document of Rio+20 (UN, 2012d).

Making it into the zero draft of the Rio+20 outcome

The timing was also propitious as SDGs figured prominently at the Department for Public Information NGO Conference that was held at the beginning of September 2011 in Bonn, Germany.[18] At this meeting, Felix Dodds, an ally of SDGs, was the Chair. The NGO Final Declaration from the Conference helped participants in presenting the Conference's recommendations on SDGs to their national governments, which were invited to contribute their inputs to Rio+20 by 1 November 2011. The Declaration proposed how to conceptualize the SDGs and outlined areas to be covered. It was interesting that the Declaration included 'Climate sustainability'[19] as one of the areas, even though Caballero at that time was cautioning about how 'very delicate climate negotiations could be adversely affected by inputs coming from outside the process but with the full weight of the UN Member States given how tenuous and highly complex these negotiations were'. She also underlined that many who were engaged in both processes felt the same. She was of the view if climate would eventually to be included as one of the SDGs, as at that time there was no certainty that even SDGs would make it into the outcome document, there would need to be a delicate balance that would not complicate or derail the ongoing UNFCCC negotiations.

The Rio+20 Regional Preparatory Meetings that were taking place in the run up to Rio+20, also referenced SDGs in their outcomes and thus increased their prominence as an accepted proposal on the table. The Latin America and the Caribbean region meeting was organized by the UN Economic Commission for Latin America and the Caribbean (ECLAC) at the beginning of September 2011. It was especially important for Caballero as she was hoping that her region would stand behind the proposal originating in one of its countries. However, even though the meeting's conclusions referenced the proposal on SDGs from Colombia and Guatemala, it stopped short of endorsing the goals. The Asia and Pacific Regional Preparatory Meeting that took place at the end of October 2011 also had in its outcome a reference to the proposals of SDGs. The regional preparatory meeting for Africa, also held at the end of October did not refer directly to the concept of SDGs in its outcome but did mention a possibility of including concrete goals for the implementation of sustainable development. The European regional preparatory meeting organized by the UN Economic Commission for Europe at the beginning of December 2011 had the strongest reference to the SDGs of all the regional meetings. However, it put them in the context of MDGs, stating that the proposal for SDGs should be drawing on the UN experience of MDGs. This was the position the EU would uphold throughout the process.

Another important boost for the SDGs around this time came from the UN Secretariat. On 21 September 2011, the UN Secretary-General, Ban Ki-Moon, speaking at the beginning of the general debate of the UN General Assembly's

66th session and launching his report 'We the peoples' put sustainable development strongly at the forefront, framing sustainable development as one of the
generational opportunities to shape the world. He characterized it as an 'imperative of the 21st century' and called for 'a new generation of sustainable development goals to pick up where the MDGs leave off' (UN, 2011). Another
staunch supporter of the SDGs as a concrete outcome of the Rio+20, was the
Secretary-General of Rio+20, Sha Zukang, a Chinese diplomat who was the
head of DESA from 2007 to 2012. He was supported by a very astute and politically savvy UN official, Nikhil Seth.[20] Then Director of the Division for Sustainable Development, Seth was very familiar with the MDGs and, as a former
Indian diplomat, was adept at reading the political situation. Thus, he recognized
early on that if the SDGs were to survive, they had to be connected to the MDGs
and above all to the post-2015 development framework. He enjoyed trust from
the heads of DESA, first Sha Zukang and later Wu Hongbo[21] as well as from the
broader UN system, especially UNDP, which was co-chairing with DESA the
UN System Task Team on the Post-2015 UN Development Agenda (UN,2012e).
This Task Team was launched by the UN Secretary-General in September 2011
to coordinate system-wide preparations for the agenda, and consisted of over 50
UN entities. Seth also enjoyed the trust of many Member States and talked often
with Caballero, helping her gauge the situation.

Despite all of these positive developments, Caballero was aware that the process was still incipient and that a key milestone was coming up. 1 November was
the deadline when submissions on the agenda of Rio+20 would need to be completed. She understood that she needed buy-in from a large number of countries
and pushed on with organizing a meeting in Bogotá.

As the 1 November deadline came, Caballero and her team were glued to the
computers. She was asking herself on that day 'Will the SDGs be in? Will enough
States put SDGs in their submissions so that they would be included in the draft?
What about international and regional organizations? Stakeholders? What if they
choose something else?'. It was a nail-biting situation, but at the end of the long
day over 50 States and international organizations, including CARICOM, mentioned SDGs as a tangible result for Rio+20 in their submissions which were to
be a base for a zero draft (UN speak for a first or early draft) for Rio+20.[22] Even
though many hurdles were still on the way to accepting the SDGs, the proposal
started gaining traction. Now delegations and stakeholders wanted to talk to
Caballero, rather than the other way round.

Refining the SDG concept

And that is why the Bogotá and later meetings in Tarrytown, New York, were
important. The SDGs were in the draft, and they started to be discussed as the
part of the outcome of Rio+20. How could Caballero make sure they *stayed* in?
'Until the last moment I was not sure whether SDGs would make it to the final
outcome document', she recalls.

The meeting in Bogotá was prior to the zero draft that was issued on 10 January 2012[23] and the Tarrytown discussions were held after. But both discussions had an important goal to achieve: solidify the position of the SDGs as a new and transformative framework for development agenda and successor of the MDGs, with two important new characteristics—universality and indivisibility of the SDGs for all nations, and a broader set of goals, not just the 'MDGs+'.

The first meeting, in Bogotá, was held on 4 and 5 November 2011. Caballero remembers it as being held in the impressive Salon Simon Bolivar, where Bolivar, the historic liberator of the country had himself worked two centuries before. The meeting was chaired by the Vice-Minister Patti Londoño that gathered over 40 delegates from a wide range of countries, including representatives from the Netherlands, Mexico, Kenya, India, Chile, the United Kingdom, Norway, the US, as well as international organizations and NGOs.

It was a positive meeting. Discussions were hard and there was a lot of push-back from those working around MDGs, both from developing countries and from the donor side. However, there was a lot of acceptance too, especially from some countries and stakeholders. 'In the end, the positive overcame the negative', recalls Dodds, who was present at the meeting. The meeting discussed how the SDGs could be used for implementation around food security. 'There was no rhetoric, little politics, and the discussion was substantive, constructive and targeted on implementation issues', Caballero recalls. This ticked all her boxes going down to details. It showed that, stripped of all politics, the SDGs were really a very good idea that would not take anything from anyone, but, on the contrary, would help improve our lives on a healthy planet.

The meeting was held under Chatham House rule,[24] so individuals felt more free to speak their minds. Still, Caballero succeeded in getting permission from the participants for a chair's summary that was tabled as 'Insights from the Informal Consultations on the SDG Proposal'. While the summary did not name people specifically, they were a good record of discussions on the scope, guiding principles, priority areas, linkages between MDGs and SDGs and the way forward. This meeting also accomplished something else: it forged friendships. It was the beginning of what came to be known as 'the SDG Friends'. As Caballero recalls,

> The SDG friends included Kitty van der Heijden of the Netherlands, Farrukh Khan of Pakistan, Jimena Leiva of Guatemala, Chris Whaley of UK, Victor Muñoz of Peru, Majid Hasan Alsuwaidi of UAE, Yeshey Dorji of Bhutan, Franz Perrez of Switzerland, Damaso Luna of Mexico, Marianne Loe of Norway, Anders Wallenberg of Sweden and many others who joined in the course of the process.

The group not only provided strong support for the SDGs, but in many hours of frustration both before and after Rio+20, was 'a space for collective catharsis' said Caballero, where people could chat freely together, airing grievances and listening sympathetically to one another.

Way to 'Rio+20'

And so, the road to SDGs continued. There was an important interlude which confirmed interest for the SDGs at the international stage. It was held in the unassuming Conference room 3 of the temporary building at UN Headquarters in New York as a special event focused on the SDGs and convened by Colombia in mid-December 2011 on the margins of the consultations on Rio+20 held by the Rio+20 Secretariat. Caballero remembers it as a short, but important event with 114 countries, because she and her team counted every one of them. 'We could still not quite believe how far we had come', said Caballero 'the momentum was growing'.

This was an important moment because it prompted Caballero to come up with a revised proposal. She sensed that after four months of consultations and especially after the meeting in Bogotá and listening to participants of the consultations for the past few days, revision was needed to reflect people's emerging views and ideas. These proposals were very political as they tried to capture the moment of where discussions were. 'I was trying to shape the process but doing so required active and engaged listening to everyone and trying to figure out what was most salient but also what could get the greatest number of people engaged', recalls Caballero.

Back at the Pod Hotel, she came with a new version. This version was much shorter and focused only on SDGs without the long explanatory introduction from the previous versions. It also contained a shorter list of focus areas for the SDGs to tackle. As always, it was supported by Vice-Minister Londoño, but this time apart from Guatemala it was also actively backed by Peru and its then President Ollanta Humala, whose support was ably presented by the then delegate of Peru to the UN Victor Muñoz.

Back in Bogotá, Caballero continued strategizing with Minister Londoño and her team. Many in the growing coalition around the SDGs felt that the meeting in Bogotá had been so rich that it was important to have another two-day international consultation but with a much broader participation in New York. And it had to take place before the formal negotiations starting in February. That was at the end of another year and only six months away from Rio+20. But Caballero's thoughts were not focused on holidays. She was scrambling to get a retreat organized in New York. Norway and the Netherlands jumped in and secured Tarrytown (40 miles north of New York City) for the meeting from 22 to 24 January 2012. The meeting was supported by the World Resource Institute (WRI), a global research organization at the nexus of environment, economic opportunity and human well-being that prepared three briefing papers for the consultations.

By that time, a zero draft of the outcome document for Rio+20 was out, and SDGs were in. Discussions were starting in earnest. Just days earlier, President Juan Manuel Santos Calderón,[25] an economist by training, had reiterated the relevance of the SDG proposal and requested the Ministry of Foreign Affairs to continue its efforts on the international level, and the Ministry of Environment and Sustainable Development to develop five national level SDGs on water, energy, food security, cities and oceans. Thus, Caballero came from her capital equipped with the highest political support from her country, while at the same time showing nationally how the SDGs were valued.

Many came to Tarrytown.[26] They gathered around a square ring of tables covered with cloth that could seat 80 in a nice warm room overlooking the snow-coated lawns of the estate. The table was long and everyone managed to fit around it. 'The fact that everyone could see each other created a sense of purpose and openness', recalls Caballero. Again, the fact that it was an informal retreat under Chatham House rule contributed to frank exchanges. A drink in the evening in a nice wood-panelled bar where delegates and other participants bonded, built trust and collegiality and, above all, a sense of ownership. She kept repeating that she wanted everyone on board and that it was not just a Colombian proposal.

And as in Bogotá, Caballero managed to get agreement to issue a Chair's summary, which provided a very good reflection of discussions and crystalized four core concepts: 'Rio+20 is a milestone event and the international community should strive for a high level of ambition, with clear and robust outcomes in the form of a renewed and focused sustainable development agenda; SDGs are understood in the context of the post-2015 development framework and have a definitive added value and will be further elaborated and completed within the post-2015 process; there should be a single unified process leading to the definition of the post-2015 framework, building upon government consultations as well as inputs from stakeholders, and expert and scientific advice; and there should be a single set of international development goals with sustainable development and poverty eradication as the overarching focus' (Colombia, 2012). It was another significant milestone, especially as it came just a day before the first negotiations on the zero draft, which started on 25 January 2012. There was still no agreement, but concepts and issues were understood by all.

Another important support to the SDGs came from the UN Secretary-General's High-Level Panel on Global Sustainability,[27] which issued its report "Resilient People, Resilient Planet: A future worth choosing" (UN, 2012e) on 30 January 2012. The Executive Secretary of the High-level Panel on Global Sustainability, János Pásztor, in presenting the report to the German Development Institute (DIE) in February, said that the 56 recommendations from the Panel report 'can be considered as a road-map of action for the negotiations in Rio in June'. Creating a set of sustainable development goals was one of its recommendations.

The Report stated that the SDGs need to be

> universal in character, covering challenges to all countries; incorporate areas that were not fully covered by MDGs such as food security, water, energy, green jobs, decent work and social inclusion, sustainable consumption and production, sustainable cities, climate change, biodiversity and

oceans, as well as disaster risk reduction and resilience; be comprehensive and cover all three dimensions of sustainable development; incorporate near-term benchmarks while being long-term in scope, looking ahead to a deadline of perhaps 2030; and engage all stakeholders in the implementation and mobilization of resources.

(UN, 2012e)

Political battle for SDGs

Now the political battles really started. There were still many who felt that sticking to the themes of Rio+20 would be the best possible option. This was a view supported by a number of developed countries, which were not in favour of reporting on their own implementation to an international body. There were those who thought that adding an environmental dimension to economic and social elements would divert resources from the 'core development' issue of poverty eradication. Many developing countries were also suspicious of the SDGs, fearing they might bring more conditionalities and restrictions from donor countries in return for their support and aid. And even though there was an increasing number of those who believed the SDGs would be the only concrete and tangible result of Rio+20, Caballero realized there was still a long way ahead and that she needed more than ever to increase her efforts to get endorsement from more countries from the Latin American and the Caribbean region, so that it would be a regional proposal for Rio+20. The latest attempt had been by the then Minister of Environment of Colombia, Frank Pearl, at the Latin American and Caribbean Environment Ministers Forum on 31 January 2012, where the region registered support, but not full endorsement, for the proposal in the Quito declaration.

And so, the outreach continued. The Colombian Mission in New York approached the Algerian Presidency of the Group of the G-77 and China in February 2012 and asked for a meeting on SDGs. Vice-Minister Londoño again played a crucial role as she personally came to New York to present the proposal at that meeting. Knowing which delegates were actually negotiating, Caballero did not request an ambassadorial level meeting, but a meeting of the delegates of the G-77 and China who dealt with these issues through the UN General Assembly's Second Committee. In other words, she wanted to get her minister in front of the people who would be doing the actual negotiating.

When Vice-Minister Londoño presented the proposal to them, delegates were impressed by the political and personal engagement at such a high level from Colombia. It was a start of many discussions in the Group on the concept and scope of the SDGs.

Negotiations on the outcome document

Activities now became focused on the negotiations of the outcome document. Colombia, through its embassy in Nairobi, Kenya, presented the proposal also to the delegates and civil society around the UN Environment Programme (UNEP) Governing Council when it met in February 2012.

Colombia also hosted another international consultation in New York in March 2012 in support of negotiations with the generous support of the Ford Foundation. This was attended by 63 countries on the margins of the negotiations on the outcome document. Support for the SDG proposal continued to be strengthened and the SDGs were further elaborated. Colombia was asked to join, as special guest, multiple events organized by the World Bank, civil society, international NGOs and others to present the proposal. Caballero remembers those events and the support she received from Swiss diplomat Franz Perrez and other countries from the SDG Friends. As the series of lunches, dinners and interminable presentations and conversations went on, she had to explain again and again the SDG concept, at the same time giving everyone a space to present their views, ask questions and even criticize the concept. Caballero believed that as long as the process was transparent and open there was a chance the concept would finally be understood for what it was and accepted.

Another important moment was at the third round of negotiations at the end of April 2012, when Caballero prepared the fourth version of SDG proposal. Now backed officially by Peru and United Arab Emirates, she worked very closely with both Victor Muñoz and Majid Hasan Alsuwaidi.

This new version of the proposal for the first time detailed the principles on which the goals should be based, stating that

> Sustainable development goals are integrated sets of voluntary, universally applicable global goal statements organized by thematic areas, with time-bound, quantitative targets and a suite (dashboard) of indicators to be adopted at national level, that aim to catalyse sound pathways to sustainable development, and to balance economic, social and environmental dimensions, and reflect the interconnections between them.
>
> *(Colombia, Peru and UAE, 2012a)*

It also showed a clear link between the MDGs and SDGs to ensure that countries who were implementing the MDGs would feel reassured that their efforts would be recognized and built upon. The paper also included a number of thematic areas for the SDGs[28] with a caveat that the list was not something proposed by the three countries but reflected 'a broad consensus around a core of issues. These issues are considered to be politically mature and to address widely acknowledged needs'.[28]

The paper continued to include another important element, which proved to be one of the most contentious in the run up to Rio+20 and during the conference itself. In fact, it was the last one to be agreed around SDGs. This element was the process of actually defining the SDGs.

Another boost to conceptualizing the SDGs as universal and interconnected came from the report that was prepared by the UN System Task Team on the

Post-2015 UN Development Agenda "Realizing the Future We Want for All". While the final version was issued shortly after Rio+20 in June 2012, an advance version was circulated before Rio+20. And even though some thought of this report as a 'Christmas tree' of issues addressed by various UN entities, it did highlight 'a vision for the future that rests on the core values of human rights, equality and sustainability' and underlined that the future agenda should be based on

> four key dimensions of a more holistic approach: (1) inclusive social development; (2) inclusive economic development; (3) environmental sustainability; and (4) peace and security" and also stressed that it "should be conceived as a truly global agenda with shared responsibilities for all countries
>
> *(UN, 2012e)*

Process of defining SDGs

As it gradually became clearer to Caballero that the SDGs had a good chance of being included in the outcome of Rio+20, she also knew it would be very difficult during the remaining couple of months to secure agreement from all countries on a clear set of goals. As a press release issued by the UN Secretariat in May 2012 acknowledged:

> Countries have differing views on what should or should not be included in the goals, as well as the formal process for how and when the goals may be defined, finalized and agreed to. Some countries would like to see the goals approved in Rio, while others see Rio+20 as a starting point for deciding on the goals. Some have concerns that the goals could bind them to commitments they feel are unrealistic, such as on climate change, while others want to ensure that countries are held accountable to achieve whatever goals are set.
>
> *(UN, 2012d)*

However, what Caballero thought was crucial to be decided at Rio+20 was a process through which SDGs would be defined *after* the Rio+20 Conference. She was convinced that only evidence- and scientifically based SDGs that were measurable would really make a difference. She had a very strong view that the

> the process to follow up a political decision in the Rio Conference needed to be focused and well structured. The components of the SDGs and the SDGs themselves must be defined through targeted consultations and deliberations by Member States, and not negotiated *prima facie*.
>
> *(Colombia, Peru and UAE, 2012a)*

In short, she did not want an overly political outcome, but one rooted in science and evidence.

Unfortunately, apart from Khan, who Caballero describes as 'a true SDG champion and an astute, seasoned and innovative negotiator who had a deep knowledge of countries' positions and an acute capacity to steer negotiations', who even though was not at the beginning supportive of the SDGs, as he admits himself, became an ally when no one else was supportive.

The G-77 and China wanted to have a typical negotiating process where New York-based delegates, often called 'the New York mafia' as they knew the inside workings at UN Headquarters, would haggle over 'particular words or commas'. This would have meant a political process, negotiated by the established groups including G-77 and China, and the EU. It would have meant a replay of the political wrangling that characterized the Rio+20 negotiations. However, the EU and other OECD countries were saying that defining the SDGs was a technical process that was best left to the Secretary-General. The rumours were also going around, which proved to be correct, that the Secretary-General would establish a high-level panel of eminent persons on the post-2015 development agenda and the then Prime Minister of the United Kingdom, David Cameron, announced that he would be co-chairing the panel.[29] Wanting to give more strength and visibility to this high-level panel, UK was one of the countries most strongly opposed to deciding on SDGs in Rio. Their officials were also arguing in favour of leaving the defining of the SDGs to after Rio+20 through a technical process to be established by the Secretary-General.

Caballero was concerned with the evolution of the negotiations. She did not want another intergovernmental document with no or little possibility to be implemented 'with empty political declarations masked as goals'. She really wanted the goals to be transformational and change the course of development for a more inclusive and sustainable world. Caballero insisted that what was being agreed should be a metric with far-reaching implications for humanity and the planet, and that it could not be bound by political considerations. She turned to all her allies throughout the SDG process and tried to convince them that unless Member States agreed on the SDGs through an open and transparent process informed by experts from the UN system and outside of it, then the SDGs would be of no real value. She envisioned a technical body of experts tackling the daunting array of issues and themes that would need to be addressed. These experts would submit a technical and, therefore, structured and evidence-based set of recommendations for approval by the UN General Assembly. She knew if she failed to convince Brazil and other Member States to have the process outlined in Rio, the SDGs would become yet another internal UN intergovernmental process that would not make a difference beyond the UN Headquarters' walls or, as was often said, 'beyond the First Avenue', where the UN was situated in New York.[30]

At the height of negotiations and with the negotiating document containing over 270 paragraphs, there was little time left for any discussion on what Caballero called 'Rio+1', that is, what would happen a day after the 'Rio+20' ended. Isabel Cavelier, a member of Caballero's team who was a climate negotiator, suggested that the Transitional Committee under the UNFCCC, a group that had successfully delivered the Green Climate Fund, be used as a model. The G-77 had wanted an intergovernmental process to establish a Green Climate Fund, while OECD countries wanted a technical process led by the UNFCCC Secretariat. At the end, the Transitional Committee consisted both of experts and Member States. It seemed a perfect fit to Caballero, so she prepared a paper with her team suggesting that a small, open, technical working group be convened based on the Transitional Committee model. It got instant support from Khan, who was a member of the Transitional Committee and felt the situation was similar. She shared the proposal informally and the idea gained traction, but again it was not smooth sailing. What worked in the UNFCCC process, as Khan later said, proved to be much more controversial in the General Assembly context.

Discussions became so heated that during the last round of negotiations in Rio itself, Khan said he left the G-77 meeting and threatened to resign as G-77 negotiator unless given free rein in this matter. His resoluteness, combined with Brazil's masterful handling of the outcome document, listening to everyone but keeping control over the written outcome, in the end resulted in agreement to create an open working group that would be transparent and work in an open space so that everyone who wanted to be present could be present.

And so, the text drafted by the two allies, Caballero and Khan, in Rio during the early hours of the morning was given to the Brazilians. As Caballero recalls, they had proposed starting with an open working group of 30 members: six from the five regional groups. However, they were sure it would be negotiated to a much higher number.

What they did not know at that time was that the Brazilians were already working on a 'take it or leave it' text that would not be open to tortuous negotiations. The Brazilians had their way and the idea of 30 members remained in the outcome document. Still, it would take seven more months after Rio+20 to negotiate the final composition of the Open Working Group on SDGs (OWG).

The SDGs become part of the Rio+20 outcome

Nevertheless, when the morning of the last day of Rio+20 dawned and the final text was proposed by the Brazilians, the concept of SDGs was in and the process of how to define them was there, too. Caballero was thrilled, standing and clapping in the faintly lit, cavernous room of the Rio Conference Centre aware that there was agreement on the concept of the SDGs and on a sensitive process to develop the framework. Her dedication of the many months had paid off.

There was a whole section in the outcome document "The Future We Want", dedicated to the SDGs. The section outlined how the SDGs would link to the MDGs, their conceptualization and incorporation into the post-2015 development agenda as well as parameters for an inclusive and transparent intergovernmental process on the SDGs that would incorporate all stakeholders and be based on science and evidence. But above all, what was important to Caballero was that it clearly stated that the SDGs would be universal and that progress towards their achievement needed to be assessed and accompanied by targets and indicators.

The reason for this favourable outcome was the need for a concrete and tangible outcome of Rio+20 on the one hand, and Caballero's thoughtful way of conceptualizing the SDGs and 'not shoving them into people's throats', as Dodds said, on the other. Caballero had given space to everyone who wanted to talk about SDGs to express their views. This helped people buy into the idea and its value. The Rio+20 outcome also led to reports and discussions linking the SDGs to the post-2015 UN development framework.

Open working group

Caballero continued to be active in the OWG (Dodds et al., 2017; Kamau et al., 2018) through 11 months of discussions during the so-called 'stock-taking phase' and all the way to May 2014, when she took up a post at the World Bank as a Senior Director, Global Practice for Environment & Natural Resources.

The OWG was by no means a usual UN negotiating process where a draft is provided and then negotiated. The two co-chairs, Macharia Kamau[31] and Csaba Kőrösi[32] showed great leadership in providing a safe space for all actors to build trust among each other, but also become knowledgeable with the issues at hand. Caballero looked at this process as a well-crafted realization of her unwavering belief that only through solid knowledge of the issues would it be possible to define the SDGs as useful tools to achieve sustainable development for all. But she

FIGURE 10.1 Paula Caballero speaking at the seventh session of the UN General Assembly (UNGA) Open Working Group on the Sustainable Development Goals, January 2014, at UN Headquarters in New York. Photograph courtesy of IISD/Earth Negotiations Bulletin | Elizabeth Press.

also believed that it should be done by Member States, not by a panel of experts that would bring them 'ready-made' to the table.

Thus, the OWG for her was a success. All voices were heard and everyone who wanted to speak spoke, even though officially the group had only 30 seats and 70 members, as seats were shared. This openness and transparency were due to the leadership of the two co-chairs but also astute steering of the process by Nikhil Seth from the Secretariat side, navigating well both the diplomatic and UN waters, being equally familiar with both. He managed to have the Secretariat and, as much as it was possible, the UN system express a unified position around the SDGs. And possibly with the quirk of destiny, he had at that time support from a team headed by David O'Connor who has an 'incorruptible scientific mind' as one of his team members described him. He was also very ably supported by the intergovernmental support branch headed by Marion Barthélemy,[33] who with her long experience at the UN, including working at the office of the Deputy-Secretary-General and Assistant-Secretary-General of DESA, possessed a keen sense of how much ambition expressed by some parts of the UN system can be presented to Member States. Thus, it was a complementarity of political astuteness embodied in Seth and Barthélemy and a complete disregard for it when it came to the substance from O'Connor, that combined with excellent leadership from the co-chairs, contributed to the final success of defining the SDGs.

This was in stark contrast to the Intergovernmental Committee of Experts on Sustainable Development Finance (ICESDF) that was mandated to be established in the Rio+20 outcome to

> assess financing needs, consider the effectiveness, consistency and synergies of existing instruments and frameworks, and evaluate additional initiatives, with a view to preparing a report proposing options on an effective sustainable development financing strategy to facilitate the mobilization of resources and their effective use in achieving sustainable development objectives.
>
> *(UN, 2012b)*

'According to many, it did not fulfil its goals and conducted its work in a completely untransparent manner closed to stakeholders. The experts nominated by their regional groups were from the more traditional "development discourse". They didn't include any of the major think tanks on sustainable development finance nor key stakeholders who were working on innovative sustainable development financial options' (Dodds et al., 2017). ICESDF is a reminder of what could have happened with the SDGs if Member States had not agreed on a different course (Dodds et al., 2017).

Meanwhile, Caballero continued to canvas, explain and even draw posters and pictures that she would put up in the OWG negotiating conference room to better illustrate her position. She tirelessly contributed to discussions on water and food systems and hunger, and participated in numerous consultations, side events, roundtables and expert meetings whenever she was invited. She stood for a dashboard approach of indicators to show how a universal agenda is possible with targets to be adapted to national circumstances. One immediate challenge that arose was a competing idea that called for two completely separate sets of goals, one for rich countries and another one for poor countries. For Caballero, such an idea was unthinkable. 'This proposal undermined some of the greatest strengths of the SDG option and would have locked countries into a static system that did not recognize the evolution of countries' development trajectories', Caballero said. She was strongly against this concept that some countries wanted and had a great support in the co-chairs who were also always emphasizing the universality of the agenda. She also believed that only by talking to everyone— delegations, UN officials, NGOs and the private sector—would it be possible to achieve a strong ownership of the goals.

It was not plain sailing. During 2013, there were still two parallel processes under way at the UN, one on SDGs through the OWG and the other one on the post-2015 development agenda. There were even two different sets of co-facilitators, one for the working group (Kamau and Kőrösi) and one for the preparations of the special event to follow up efforts made towards achieving the MDGs that took place on 25 September 2013 (Permanent Representatives of Ireland and South Africa).[34] However, the Member States were becoming uneasy with this parallel process, no one more so than Caballero. She had from the beginning been working to unify the two parallel agendas saying that there could and should not be a separate environment agenda if transformational changes were to happen and if the world should be put on the path to sustainable development. She did her part, trying to exert influence through Colombia's participation in the High-level Panel of Eminent Persons on the Post-2015 Development Process, where she sought to include the Rio+20 process alongside the Millennium Declaration (Dodds et al., 2017).

Unfortunately, the then Foreign Minister María Ángela Holguín was unable to attend the first meeting that took place in London on 2 November 2012. Caballero, who was her sherpa (or advisor) was there, but only Panel members were allowed to speak. She was concerned that issues around sustainability were not getting sufficient attention, so she wrote a paper that the Minister approved. It was circulated ahead of the second day's meeting of the Panel. This resonated with the Panel's leadership and at the following meeting in Liberia that took place on 1 February 2013, Colombia's foreign minister was invited to give a presentation on sustainable consumption and production. In preparing it, Caballero was supported by WWF and WRI. Minister Holguín made a compelling case for a circular economy and sustainable and transparent management of natural resources. This put strongly the three dimensions of sustainable development on the table and the issuing framing questions by the Panel, the two of which read 'What is the role of the SDGs in a broader post-2015 framework?' and 'Should (social, economic and environmental) drivers and enablers of poverty reduction and sustainable development, such as components of inclusive growth, also be included as goals?' included the three dimensions of sustainable development.[35] And the final communique from the meeting stated that 'This is a global, people-centered and planet-sensitive agenda to address the universal challenges of the 21st century: promoting sustainable development, supporting job-creating growth, protecting the environment and providing peace, security, justice, freedom and equity at all levels' (UN, 2013a).

Caballero felt that overall the report from the High-level panel, which contributed to the report of the Secretary General (UN, 2013c)[36] for the special event in September 2013, was positive. The Panel helped shape the playing field for the discussions on the SDGs, bringing in key concepts such as that of the concept of leaving no one behind, an idea that had strong inputs from the secretary of the High-level panel Homi Kharas from the Brookings Institution.

This view of having one agenda was also reflected in the outcome document of the special event which was adopted by the General Assembly (UN, 2013b) stipulating that 'We are resolved that the post-2015 development agenda should reinforce the commitment of the international community to poverty eradication and sustainable development' (UN, 2013b). The outcome also launched the negotiations on the post-2015 development agenda, stating 'We decide today to launch a process of intergovernmental negotiations at the beginning of the sixty-ninth session of the General Assembly, which will lead to the adoption of the post-2015 development agenda' (UN, 2013b).

An important role in the work of the High-level panel and later in OWG and post-2015 development agenda was played by Amina Mohammed.[37] She was an *ex-officio* member and worked tirelessly to connect all the streams and build support for the SDGs. She stayed in this role until 2015 and advised the UN Secretary-General, acting as his link to the OWG. Mohammed became an eloquent spokesperson for the SDGs, and the idea of just revamping the MDGs with a little bit of window dressing—which appeared to be the original plan for the post-2015 agenda—fell by the wayside. She said that

> the MDGs have helped to end poverty for some, but not for all. The SDGs must finish the job and leave no one behind. The SDGs must help us build the future we want, a future free from poverty and one that is built on human rights, equality and sustainability.
>
> *(Dodds et al., 2017; Kamau et al., 2018)*

With all these developments, Caballero knew that the SDGs and the goal of sustainable development were now on track to be adopted and that all her efforts were bearing fruit.

The OWG carried out its task under the able leadership of the two co-chairs supported by the UN Secretariat harboured in DESA and came to the consensual outcome of 17 SDGs that it forwarded to the General Assembly for consideration in the context of the post-2015 development agenda process (Kamau et al., 2018).

While there were attempts to reopen and relitigate the outcomes of the OWG, the G-77 and China stood firm and did not let it happen—something that Caballero welcomed (Dodds et al., 2017; Kamau et al., 2018). By this time, many people felt great ownership of the SDG idea and understood that a consensus achieved was precarious, and any tampering with it would unravel the whole document. The Sustainable Development Summit on 27 September 2015 adopted 'Transforming our world: the 2030 Agenda for Sustainable Development' by acclamation from all 193 Member States of the United Nations. The SDGs reflected the agreement in the OWG, except for some minor technical tweaks for some targets.

What are the SDGs about?

The SDGs are a set of 17 interlinked goals and 169 related targets designed to be a blueprint to achieve a better and more sustainable future for all. They are a call for action by all countries—developed and developing—to promote prosperity while protecting the planet. In the face of a pandemic, they are providing a framework for sustainable recovery where benefits will be more equally shared, and societies will be more inclusive providing opportunities for all and nature and the environment will be protected and natural resources managed sustainably and responsibly. They represent the whole complexity of development, but also a high ambition to achieve a better life for all now. SDGs need to be achieved in their entirety and no SDG is more important than the other and they are all interlinked, so progress in one SDG will benefit other SDGs. This requires unprecedented coordination and cooperation by the whole-of-government and the whole-of-society where literally no one must be left behind.

What the future holds

When asked about the future and what needs to be done to make SDGs a reality for all, Caballero has only two words: 'transformation' and 'interlinkages'. She believes the SDGs are not a silver bullet, but that a world without them would be much worse off. The SDGs created a common language and a discourse of an integrated, global development. Even though the earlier MDGs at one point played an important role for developing countries, that agenda was limited and did not encompass the holistic view of human life and planetary systems. It also represented only the basic needs of any human being, while leaving ideas of a more prosperous life for somewhere in the future. The SDGs, however, for the first time in the history of development represent a way to achieve a good quality of life for all within the planetary boundaries right now. And for this to be achieved, as Caballero says, cannot be done by 'MDG mentality'.

Caballero understood the complexity of development and the predicament that politicians face. For her the only solution was to have integrated approaches so that gains in one sector or area could benefit others and thus minimize trade-offs. She knew that times are difficult in an increasingly polarized world. She also knows that a lot still needs to be done to improve how we implement the SDGs.

The world today is at the crossroads. There is increasing inequality within and between societies with billions left behind and overwhelming evidence of rising global risks due to ever increasing human pressures on the planet. There are only nine more years to achieve the SDGs and the progress has been slow, further slowed by the COVID-19 pandemic. Between 119 and 124 million more people fell into extreme poverty in 2020, further compounding challenges to poverty eradication such as conflict, climate change and natural disasters. The crisis is also exacerbating inequalities: in 2020, the equivalent of 255 million full-time jobs were lost. The climate crisis is still worsening with the global average temperature about 1.2°C above pre-industrial levels in 2020, dangerously close to the 1.5°C limit established in the Paris Agreement.

Without deep transformational changes, we won't be able to achieve a sustainable, just and resilient future for all. And Caballero constantly underlines that what is measured needs to be policy relevant and that all policies and strategies need to be brought together to achieve better and prosperous lives. For her, the SDGs need to be implemented through the Paris Agreement on climate change (UNFCCC, 2015)[38] lens and vice versa.

And what is Caballero doing today? In 2014, she left the Colombian Government to work first in the World Bank, then in the WRI and then at Rare, a US-based environmental organization where she focused on natural resources, land, water and climate. In 2021, she took a new position as the Regional Managing Director for Latin America in The Nature Conservancy, where she will be implementing what she has been advocating for so long.

While we face many global challenges today, the world has already been positively affected because of Caballero and many more like her who have embraced the idea of well-being for all human beings wherever they are on this Earth and the need for humans to live in harmony with nature.

Notes

1 The economic and social committee of the UN General Assembly is one of the six main committees of the General Assembly that is open to all 193 Member States of the United Nations. It is referred to among delegates as the 'Second Committee'. This committee deals with all the outcomes of conferences in economic, social and related fields and the General Assembly resolutions related to sustainable development were first negotiated in the Second Committee and then adopted at the plenary of the General Assembly. The resolution 64/236 was adopted in the Second Committee and then in the General Assembly on 24 December 2009.

2 The United Nations Commission on Sustainable Development (CSD) was established by the UN General Assembly in December 1992 to ensure effective follow-up of the United Nations Conference on Environment and Development (Rio, 1992). It held its last session (CSD-20) on Friday, 20 September 2013 prior to the meeting of the United Nations High-level political forum on sustainable development (HLPF) as decided in the UN General Assembly resolution 67/203. HLPF has replaced CSD.

3 The name came because of the first Rio Conference on Environment and Development which resulted in the Agenda 21 that happened in 1992 in Rio, 20 years before the second Rio.

4 Brazil hosted the Conference on Environment and Development in 1992.

5 South Africa hosted the World Summit on Sustainable Development in 2002 which resulted in the Johannesburg Plan of Implementation.

6 Speech at the general debate of the UN General Assembly's 62nd session on 25 September 2007.

7 The Group of 77 (G-77) was established on 15 June 1964 by 77 developing countries signatories of the 'Joint Declaration of the Seventy-Seven Developing Countries' issued at the end of the first session of the United Nations Conference on Trade and Development (UNCTAD) in Geneva. The first 'Ministerial Meeting of the Group of 77 was in Algiers (Algeria) on 10–25 October 1967, which adopted the Charter of Algiers'. Since then the group increased its membership to 134, but the name was retained due to its historic significance. The Group of 77 says for itself that 'it is the largest intergovernmental organization of developing countries in the United Nations, which provides the means for the countries of the South to articulate and promote their collective economic interests and enhance their joint negotiating capacity on all major international economic issues within the United Nations system, and promote South-South cooperation for development'. The Chairman, who acts as its spokesman, coordinates the Group's action in each Chapter. The Chairmanship, which is the highest political body within the organizational structure of the Group of 77, rotates on a regional basis (between Africa, Asia-Pacific and Latin America and the Caribbean) and is held for one year. Currently the Republic of Guinea holds the Chairmanship of the Group of 77 in New York for the year 2021. The Annual Meeting of the Ministers for Foreign Affairs of the Group of 77 is convened at the beginning of the regular session of the General Assembly of the United Nations in New York and at this meeting usually the Chair for the following year is selected. China, who even though provides political and financial support to the group is listed separately as it does not consider itself as a member and so the group is referred to as 'G-77 and China'.

8 View expressed in the interview by Farrukh Khan, August 2021. Farrukh Khan, Ambassador was Chef-de-Cabinet at the 75th session of the United Nations General Assembly since 21 June 2021 until the end of the session. Prior to this, Farrukh was Deputy-Chef-de Cabinet at the 75th session and also has over 25 years of varied diplomatic experience with the Foreign Service of Pakistan and at the UN as a Senior Manager in the Executive Office of the Secretary General. He was also Counsellor at the Permanent Mission of Pakistan to the UN in New York and main coordinator for Rio+20 for G-77 and China and a member of the Bureau of the UN Conference on Sustainable Development (UNCSD).

9 All quotes throughout the chapter are from interviews conducted with Paula Caballero by the author in August and September 2021.

10 Millennium Development Goals that came out of the Millennium Declaration in 2000. There were eight MDGs: (1) Eradicate extreme poverty and hunger; (2) Achieve universal primary education; (3) Promote gender equality and empower women; (4) Reduce child mortality; (5) Improve maternal health; (6) Combat HIV/AIDS, malaria and other diseases; (7) Ensure environmental sustainability; and (8) Global partnership for development. They were never intergovernmentally agreed and were conceptualized by the Secretary-General and the UN system.

11 United Nations Conference on the Environment, 5–16 June 1972, Stockholm, Sweden. This was the first world conference to make the environment a major issue. It started a dialogue between developed and developing countries on the link between economic growth, the pollution of the air, water and oceans and the well-being of people around the world. The major outcome of this conference was the creation of the United Nations Environment Programme (UNEP).

12 Poverty Reduction Strategy Papers (PRSPs) were documents required by the International Monetary Fund (IMF) and World Bank before a country could be considered for debt relief within the Heavily Indebted Poor Countries (HIPC) initiative. PRSPs were also required before low-income countries could receive aid from most major donors and lenders. They were primarily focused on economic growth and reducing poverty through macroeconomic measures.

13 David O'Connor headed policy and analysis branch in the Division for sustainable development in UN DESA from 2004 to 2016 and is currently Permanent Observer of IUCN to the United Nations in New York.

14 Agenda 21 formalized nine sectors of society as the main channels through which broad participation would be facilitated in UN activities related to sustainable development. These are officially called 'Major Groups' and include the following sectors: women, children and youth, indigenous peoples, non-governmental organizations, local authorities, workers and trade unions, business and industry, science and technological community and farmers.

15 Interview in August 2021.

16 Felix Dodds is a sustainable development consultant and adjunct professor at Environmental Sciences and Engineering and Senior Fellow, Global Research Institute, University of North Carolina and Associate Fellow, Tellus Institute, Boston and author of over 20 books.

17 United Nations is divided in five regional groups which are used for all elections and steering bodies where all of these groups have to be represented. They are African States/African Group, Asia-Pacific Group, Eastern European States Group, Group of Latin American and Caribbean States (GRULAC) and Group of Western European and Other States (WEOG). There regions are also used for regional preparatory meetings as was the case in the run up to Rio+20.

18 64th Annual United Nations Department of Public Information/NGO Conference was held from 3 to 5 September 2011 in Bonn, Germany under the title 'Sustainable Societies; Responsive Citizens'.

19 'SDG-CLIMATE SUSTAINABILITY: By 2050, governments should have reached clear pathways towards climate sustainability that regulates the global temperature rise below 1.5 degrees C. Emissions of greenhouse 481 gases should be reduced to 25% of 1990 levels by 2020, 40% by 2030, 60% by 2040 and 80% by 2050. Carbon taxes and tariffs should be in place to provide incentives for low-carbon development and manufacturing, finance GHG emissions reduction projects, REDD+ and other offset mechanisms, and green infrastructure solutions to help vulnerable communities adapt to climate change', Declaration of the 64th UN DPI/NGO Conference, Bonn, Germany, 3–5 September 2011. Available online at: https://www.un.org/sites/un2.un.org/files/64th_un_dpi_ngo_conference_ger_-_final_report.pdf.pdf

20 Nikhil Seth was a delegate to the UN in the Permanent Mission of India to the United Nations from 1990 to 1993 and negotiator in the first Rio conference in 1992. He has held various positions within the UN Secretariat from 1993 to date. He headed the Division for sustainable development in UN DESA from 2011 to 2015 which acted as the Secretariat for the Rio+20 Conference, OWG and post-2015 development agenda. He currently heads the UN Training and Research Institute (UNITAR) in Geneva as Assistant-Secretary General.

21 Wu Hongbo is a Chinese diplomat who headed DESA from August 2012 to August 2017.

22 Zero draft is an expression used for negotiating texts at the United Nations that does not yet have a status. It is a draft usually prepared by the co-facilitators, co-chairs, etc. of the process, in this case co-chairs of the UNCSD Bureau the then Permanent Representatives of Antigua and Barbuda, H.E. Mr. John Ashe and the Republic of Korea, H.E. Mr. Kim Sook.

23 The negotiations on the zero draft, which was based on over 6,000 pages of inputs received by the UNCSD Secretariat from Member States, UN system, other

        international and regional organizations as well as major groups and other stakehold-
ers were staggered. The first two sections: Section I, 'Preamble/Stage setting', and
Section II, 'Renewing Political Commitment' were discussed during the intersessional meeting from 25 to 27 January 2012 and the rest (Section III, 'Green economy
in the context of sustainable development and poverty eradication'; Section IV, 'Institutional framework for sustainable development'; and Section V, 'Framework for
action and follow-up' where SDGs were included) was discussed at another informal
meeting from 19 to 23 March 2012.

24 Chatham House rule is 'When a meeting, or part thereof, is held under the Chatham
House Rule, participants are free to use the information received, but neither the identity
nor the affiliation of the speaker(s), nor that of any other participant, may be revealed'.

25 Juan Manuel Santos Calderón was President of Colombia from 2010 to 2017. He is
a recipient of the 2016 Nobel Peace Prize for his efforts to bring the country's more
than 50-year-long civil war to an end. Being an economist, he understood well the
value of SDGs.

26 Forty-four countries and many UN representatives including the Secretary-General of
the Conference, the Executive Coordinators of Río + 20 and representatives of the UN
system as well as NGOs and civil society were part of the retreat in Tarrytown, NY.

27 United Nations Secretary-General Ban Ki-moon launched a 21-member High-Level
Panel on Global Sustainability on 9 August 2010 co-chaired by the Finnish President
Tarja Halonen and South African President Jacob Zuma. The panel included high-
level officials from Australia, Barbados, Canada, China, European Union Commis-
sioner for Climate Action, India, Japan, Mexico, Mozambique, Nigeria, Republic
of Korea, Russian Federation, Spain, Sweden, Switzerland, Turkey, UAE, USA and
Gro Harlem Brundtland, Former Prime Minister of Norway and former Chair of the
World Commission on Environment and Development whose Commission came up
with the most widely accepted definition of sustainable development as development
that meets the needs of the present, without compromising the ability of future generations to meet their own needs. Report of the World Commission on Environment
and Development "Our Common Future" (1987).

28 The list included areas proposed by Johannesburg Plan of Implementation from the
World Summit on Sustainable Development, Secretary-General of the Conference,
zero draft as well as by major groups and other stakeholders and those were: poverty,
sustainable consumption and production, food, water, energy cities, oceans, growth
and jobs, natural resources, gender, climate change, technology, education, disasters,
biodiversity, land degradation and forests.

29 The UN Secretary-General Ban Ki-moon announced the 27 members of a High-
level Panel to advise on the global development framework beyond 2015 in July
2012 co-chaired by President Susilo Bambang Yudhoyono of Indonesia; President
Ellen Johnson Sirleaf of Liberia; and Prime Minister David Cameron of the United
Kingdom. The panel consisted of representatives from governments of Benin, Brazil,
China, Colombia, Cuba, France, Japan, Jordan, Latvia, Mexico, Nigeria, Russian
Federation and Timor-Leste and representatives of academia, civil society and private
sector from, India, Kenya the Netherlands, South Africa, Turkey, USA and Yemen.

30 UN Headquarters in New York are situated in the Turtle Bay neighbourhood of
Manhattan, on 17–18 acres of grounds overlooking the East River. Its borders are
First Avenue on the west, East 42nd Street to the south, East 48th Street on the north
and the East River to the east.

31 Macharia Kamau was the Permanent Representative of Kenya to the United Nations
in New York from 2010 to 2018 and after successfully spearheading the OWG, became one of the co-chairs for the post-2015 development agenda. He is currently the
Principal Secretary to the Ministry of Foreign Affairs of Kenya.

32 Csaba Kőrösi was the Permanent Representative to the United Nations in New
York from 2010 to 2014 and is currently Head of Directorate for Environmental

Sustainability in the Ministry of Foreign Affairs of Hungary and advisor to the President of Hungary.

33 Marion Barthélemy is currently Director of the Office of Intergovernmental Support and Coordination for Sustainable Development of the United Nations Department of Economic and Social Affairs (DESA). From 2012 to 2016, she served as Chief of the Intergovernmental Support Branch at the Division for Sustainable Development at the same Department. From 2000 to 2005, she served in the Office of two Under-Secretary-Generals for Economic and Social Affairs, supported the Assistant Secretary-General for Policy Coordination and served as Special Assistant to the UN Deputy Secretary-General.

34 Ambassador David Donoghue was Permanent Representative of Ireland to the United Nations in New York from 2013 to 2017 and was a co-facilitator of the post-2015 development agenda. Ambassador Jeremiah Kingsley Mamabolo was Permanent Representative of South Africa to the United Nations in New York from 2013 to 2016.

35 Framing questions agreed by the High-level panel of eminent persons on the post-2015 development agenda. Available online at: https://www.un.org/sg/sites/www.un.org.sg/files/documents/management/HLP_Framing_Questions.pdf

36 UN Secretary General, Ban-Ki Moon presented a report "A life of dignity for all: accelerating progress towards the Millennium Development Goals and advancing the United Nations development agenda beyond 2015" at the special event which drew on inputs from the High-level Panel of Eminent Persons on the Post-2015 Development Agenda, United Nations Development Group consultations, the Global Compact and the Sustainable Development Solutions Network.

37 Amina Mohammed has been UN Deputy-Secretary-General since 2017. She was appointed by the UN Secretary-General Ban Ki-Moon in June 2012 as his special adviser for post-2015 development planning and finished her mandate in 2015.

38 Paris agreement on climate change is a legally binding international treaty on climate change. It was adopted by 196 Parties at COP 21 in Paris, on 12 December 2015 and entered into force on 4 November 2016. Its goal is to limit global warming to well below 2, preferably to 1.5 degrees Celsius, compared to pre-industrial levels.

References

Colombia (2011) RIO+20: Sustainable Development Goals (SDGs): A Proposal of January 2011.

Colombia (2012) Retreat on Sustainable Development Goals, Rio+20 and the Post-2015 Development Agenda, Colombia (2012) Chair's Summary Tarrytown, NY, 23–24 January 2012.

Colombia and Guatemala (2011) RIO+20: Sustainable Development Goals (SDGs) A Proposal of March 2011.

Colombia, Peru and UAE (2012a) Proposal on SDGs of April 2012.

Colombia, Peru and UAE (2012b) Updated Proposal on SDGs of May 2012.

Dodds, F., Ambassador Donoghue, D. and Leiva Roesch, J. (2017) Negotiating the Sustainable Development Goals: A Transformational Agenda for an Insecure World, Routledge.

Kamau, M., Chasek, P. and O'Connor, D. (2018) Transforming Multilateral Diplomacy, Routledge.

United Nations (2008) United Nations General Assembly Resolution 63/212 of 19 December 2008, UN. Available online at: https://undocs.org/en/A/RES/63/212

United Nations (2009) United Nations General Assembly Resolution 64/236 of 24 December, 2009, UN. Available online at: https://undocs.org/en/A/RES/64/236

United Nations (2012a) Current Ideas on Sustainable Development Goals and Indicators, Policy Brief 6, UN. Available on line at: https://sustainabledevelopment.un.org/content/documents/327brief6.pdf

United Nations (2012b) Back to Our Common Future Sustainable Development in the 21st Century (SD21) Project, Summary for Policymakers, UN. Available online at: https://sustainabledevelopment.un.org/content/documents/UN-DESA_Back_Common_Future_En.pdf

United Nations (2012c) The Future We Want United Nations General Assembly Resolution 66/288 of 27 July 2012, UN. Available online at: https://www.un.org/ga/search/view_doc.asp?symbol=A/RES/66/288&Lang=E

United Nations (2012d) Press Release of 4 May 2012.

United Nations High-Level Panel on Global Sustainability (2012e) Resilient People, Resilient Planet: A Future Worth Choosing, UN. Available online at: https://www.uscib.org/docs/GSPReportOverview_A4 size.pdf

United Nations (2013a) Communiqué from the Meeting of the High-Level Panel of Eminent Persons on the Post-2015 Development Agenda in Monrovia, Liberia of 1 February 2013, UN. Available online at: https://www.un.org/sg/sites/www.un.org.sg/files/documents/management/Monrovia_Communique_1_Feb_2013.pdf

United Nations (2013b) Communiqué from the Meeting of the High-Level Panel of Eminent Persons on the Post-2015 Development Agenda in Bali, Indonesia of 27 March 2013. Available online at: https://www.un.org/sg/sites/www.un.org.sg/files/documents/management/Final Communique Bali.pdf

United Nations (2013c) Outcome Document of the Special Event to Follow Up Efforts Made Towards Achieving the Millennium Development Goals United Nations General Assembly Resolution 68/6 of 9 October 2013, UN. Available online at: https://undocs.org/en/A/RES/68/6

United Nations (2015) The 2030 Agenda for Sustainable Development the United Nations General Assembly Resolution 70/1 of 25 September 2015. Available online at: https://www.un.org/ga/search/view_doc.asp?symbol=A/RES/70/1&Lang=E

UN Framework Convention on Climate Change – UNFCCC (2015) Paris Agreement on Climate Change. Available online at: https://unfccc.int/sites/default/files/english_paris_agreement.pdf

United Nations Secretary-General (2011a) Accelerating Progress Towards the Millennium Development Goals: Options for Sustained and Inclusive Growth and Issues for Advancing the United Nations Development Agenda Beyond 2015, Annual Report of July 2011, A/66/126. Available online at: https://undocs.org/A/66/126

United Nations Secretary-General (2011b) Speech at the Launch of His Report "We the Peoples" of 21 September 2011, UN. Available online at: https://www.un.org/sg/en/content/sg/speeches/2011-09-21/address-66th-general-assembly-we-peoples

United Nations Secretary-General (2013) A Life of Dignity for All: Accelerating Progress Towards the Millennium Development Goals and Advancing the United Nations Development Agenda Beyond 2015, UN. Available online at: https://undocs.org/A/68/202

United Nations Secretary-General (2021) Progress Towards the Sustainable Development Goals, Annual Report March 2021. Available online at N2110971.pdf (un.org)

UN System Task Team on the Post-2015 UN Development Agenda (UN, 2012e) Realizing the Future We Want for All, UN. Available online at: https://www.un.org/en/development/desa/policy/untaskteam_undf/untt_report.pdf

11

CONCLUSION

Chris Spence and Felix Dodds

This book is about environmental heroes: exceptional people facing critical issues the world needs to address.

It tells the stories of a dozen individuals who have played a global role in taking on an environmental challenge that is larger than their organization or their country. None of them were superheroes imbued with special powers. They were just regular people like you and me. And yet, each of them made a commitment to making the world a better place. They were determined to do something in their sphere of influence, their area of knowledge or expertise, to help, no matter how hard or long the journey. Because of their courage and determination, they have improved our world.

In a time when we seem to be lurching from crisis to crisis, this book shows we should never lose hope. Every story told here illustrates how we can, as individuals working with others, make a material difference.

Did each person in our book 'save the world'? No. Today we still face many global challenges, some of them dauntingly large. But without these heroes, the challenges would be even greater.

For instance, if Mostafa Tolba and his allies had not helped establish the Montreal Protocol, tens of millions more people would already have had skin cancer, and tens of thousands more would have died. Instead, the ozone layer is healing.

Without Sidney Holt, many more cetacean species may have become extinct, these wonderful creatures consigned to history books. Instead, many are recovering. And heroes like Maurice Strong, Maria Luiza Viotti and Paula Caballero have advanced the idea of sustainable development into the mainstream of our global conversation—a critical achievement.

DOI: 10.4324/9781003202745-12

This, then, is the major conclusion of our book:

that people need to step up and lead. They have in the past, as this book shows. And they can—and must—in the future.

This book also shows that multilateral diplomacy, when done right, actually works.

The idea of multilateralism—of governments and other stakeholders working together at a global level—has been criticized by many politicians of various political stripes over the past 20 years or more. International negotiations are complex and messy, they say. It can take years of effort to achieve anything. It also means governments sometimes have to give up some control, even some sovereignty. And it can fail, say the critics, leaving us back where we started.

All of this is true.

But it is also undeniably the only way to solve global problems. No one, not the US or China or the EU or anyone else, can solve climate change by themselves. It's just too big. The same goes for our goals of achieving sustainable development, protecting biodiversity, reducing the global risks of chemicals and hazardous wastes, maintaining our fisheries, preventing deforestation or any of a host of other problems.

The truth is, we need each other. We need to work together. Yes, it will always be messy and complex and frustrating and exasperating and often risky. But this book shows that just about anything is possible when it's done right: when it's led by individuals with vision and determination and backed by governments and other stakeholders willing to invest in working together for a common cause.

Why these issues?

Any book of this kind is selective. We couldn't include every environmental topic that has an international profile. Instead, we have tried to tell stories of remarkable people who had an impact in key areas. On many topics, these stories are not yet over. In fact, most of the challenges we reflect on in the book not only remain, but are growing.

Take climate change, for instance. Why did we devote three chapters of our book to this issue when the climate situation is so dire? There were two main reasons. First, we wanted to show progress that has already been made. Have we solved the problem? Far from it. But we believe that without the Kyoto Protocol, the Paris Agreement and even the much-maligned Copenhagen Accord, the world would be in a far worse situation than it is now. These international agreements prompted us to start changing how we have been running our economies for the past 200 years. Progress has been agonizingly slow at times. Nevertheless, Kyoto, Copenhagen and Paris have helped create not only global awareness, they have given us a roadmap to move forward. This has triggered rapid improvement and deployment of new green technologies such as wind and solar. They have also generated a growing acceptance that we need to limit temperature rises to just 1.5°C or less. Without these agreements, it would probably be already too late to avoid catastrophic climate change. Because of them, we still stand a chance.

The second reason for including these chapters on climate change is that we wanted to celebrate successes that are often forgotten in the shadows of our current doom-and-gloom moment. We want to illuminate them not because we believe the job is done (it isn't), but in the hope they inspire future heroes and show these challenges can be overcome.

For our other topics, we wished to show the breadth of environmental challenges as well as the international community's response. Protecting nature and combating biodiversity loss are enormous issues. The Ramsar Convention and International Whaling Commission provide good examples of early, committed action.

The damage caused by human-made chemicals is another global problem. Here, the Montreal Protocol on the ozone layer and the Basel Convention and the Ban Amendment on the international movement of hazardous waste show what can be done with dedication and strong, science-based evidence.

Sustainable development remains our ultimate goal, and the chapters on the Rio Earth Summit, Rio+20 and the Sustainable Development Goals illustrate what we've achieved in this direction so far.

What and who's missing?

There are many other treaties and processes we could have included but did not, for a variety of reasons. In many cases, we did not feel the story was ready to tell, either because we believe more progress is needed or the timing just isn't quite right. They may form the basis of a second book sometime in the future.

As for the people we featured, we want to emphasize that success is never down to one person. Every story featured in this book required many people to bring it about. It is always a collective, team effort, even if individual roles are highlighted here.

So who else could have featured as one of the heroes in this book? For a start, Johannah Bernstein, Pamela Chasek and Langston "Kimo" James Goree VI could make a strong claim. They set up the *Earth Negotiations Bulletin* (*ENB*), a publication that has reported from almost every single major environmental negotiation since the early 1990s. The ENB casts light on what is happening behind closed doors in negotiations, creating much needed transparency and accountability on why certain meetings succeed while others do not. Kimo Goree led *ENB* with dedication and passion for more than a quarter of a century; Pamela Chasek remains its editor to this day. Their impact, and that of the teams of experts who attended meetings for ENB over the past 30 years, is incalculable.

Chip Lindner is another who, like those at the ENB, played a critical role in providing information for stakeholders to engage in these global processes. *The Centre for Our Common Future* with its *Network* newsletter set up a monthly update for everyone preparing for the Earth Summit in Rio in 1992 and hosted the first real Global Forum around a multilateral negotiation. This has been copied by many people and organizations ever since as a blueprint for how things should work.

Then there are many politicians from all stripes who have made a difference, not to mention scientists, diplomats and stakeholder activists. For those who fully deserve to be featured in the pages of such a book as this, we apologize that we could feature only a small number of the many heroes of environmental diplomacy. Perhaps a future book might tell more of these stories?

And what might a future edition look like? We know this book is a product of its time. With stories ranging from the 1960s through to the 2020s, it may not be a surprise that most of the chapters—seven of the ten—are about men. We believe—and firmly hope—that were this book to be updated in 5 or 10 or even 20 years, more of the heroes would be women. Indeed, it is interesting to note that, of the four stories set after 2010, three of our four heroes *are* women.

The book also represents a geographically diverse group of people. Again, because the world was a different place when our stories begin, those set in the last century tend to focus more on Western men. Even so, we have an interesting cast of characters, with heroes from Argentina, Brazil, Canada, Costa Rica, Colombia, Egypt, Iran, Switzerland (2), the UK (2) and the US. It is worth noting that by 2030, seven of the largest ten countries by GDP will be what is considered today to be developing countries. Our guess is that a future edition set, say, a decade from now, would include more heroes from Asia, the Pacific, Latin America, the Caribbean and Africa. Still, even within the Western-dominated world in which much of this book is set, it is interesting to see protagonists from every continent except Australia and Antarctica. We believe this shows heroes can come from anywhere.

And make no mistake, we will need more heroes. While this book has demonstrated that leaders can come from anywhere and be completely different in personality or background, they do all share one thing: persistence. In the years to come, we will need this quality more than ever if we are to solve the myriad global challenges that confront us today.

Finally, we wish to end on a note of hope. History shows us that even things that seem unattainable can be achieved. As the great Nelson Mandela said:

"It always seems impossible until it's done."

INDEX

Note: *Italic* page numbers refer to figures and page numbers followed by "n" denote endnotes.

Printed in the United States
by Baker & Taylor Publisher Services